ROUTLEDGE LIBRARY EDITIONS:
EXCHANGE RATE ECONOMICS

Volume 4

THE INTEREST STANDARD OF CURRENCY

THE INTEREST STANDARD OF CURRENCY

An Attempt

ERNST DICK

Routledge
Taylor & Francis Group
LONDON AND NEW YORK

First published in 1925 by George Allen & Unwin

This edition first published in 2017
by Routledge
2 Park Square, Milton Park, Abingdon, Oxon OX14 4RN

and by Routledge
605 Third Avenue, New York, NY 10017

Routledge is an imprint of the Taylor & Francis Group, an informa business

British Library Cataloguing in Publication Data
A catalogue record for this book is available from the British Library

ISBN: 978-0-415-79325-4 (Set)
ISBN: 978-1-315-21117-6 (Set) (ebk)
ISBN: 978-1-138-63063-5 (Volume 4) (hbk)
ISBN: 978-1-315-20929-6 (Volume 4) (ebk)

THE INTEREST STANDARD OF CURRENCY: AN ATTEMPT

BY

ERNST DICK, PH.D.

LONDON: GEORGE ALLEN & UNWIN LTD.
RUSKIN HOUSE, 40 MUSEUM STREET, W.C. 1

First published in 1925

Printed in Great Britain by
UNWIN BROTHERS, LIMITED, LONDON AND WOKING

PROLOGUE

WHAT OUR FATHERS KNEW

ALL variations in the value of the circulating medium are mischievous: they disturb existing contracts and expectations, and the liability to such changes renders every pecuniary engagement of long date entirely precarious. The person who buys for himself, or gives to another, an annuity of £100, does not know whether it will be equivalent to £200 or to £50 a few years hence. Great as this evil would be if it depended only on accident, it is still greater when placed at the arbitrary disposal of an individual or a body of individuals; who may have any kind or degree of interest to be served by an artificial fluctuation of fortunes; and who have at any rate a strong interest in issuing as much as possible, each issue being in itself a source of profit. Not to add, that the issuers may have, and in the case of a Government paper always have, a direct interest in lowering the value of the currency, because it is the medium in which their own debts are computed.—JOHN STUART MILL, *Principles of Political Economy*, Book III, chapter xiii, § 1.

PREFACE

This book sets forth a new scheme to stabilize the currency. It may seem to appear somewhat late in the day; this talk of stabilization has grown rather stale—we have heard enough of it. Besides, the world is thoroughly weary of reforms, disenchanted and sceptical. It is in a state of sensitiveness that makes it turn in a sort of anguish from anything implying change or effort. Well, it so happens that this state fits in admirably with the requirements of my scheme. When I speak of stabilization I mean that something ought to cease from squirming. And since every something is bound up with every other thing, my demand of stabilization includes the necessity of leaving things alone generally. Whereas the reform schemes that have preceded this late-comer of mine agree in demolishing certain time-honoured institutions and in denouncing a certain state of things—not otherwise to be got at and damaged—my plan demands that everything should be left where we find it at the present moment. There may be more substance in this demand than a first hearing of the proposal seems to suggest. "At the present moment" (say, the beginning of 1925 in the countries which have tolerably well recovered from the war, such as England, the United States of America, and my own country of Switzerland) signifies a final breach with the so-called "normal" past of ante-war conditions: no return to anything in particular! No return to the gold standard in the form of an officially fixed price of gold; no fixed ratio of the note issue to the gold reserve; no return to a discount policy intended to regulate the flow of gold. The perpetuation of the present state of things implies these three main points:

(1) Leave the price of gold to find its own level, but do not debar gold from being freely used as legal tender, nor from moving freely from one country to another.

(2) Leave the rate of discount unchanged. Both England and Switzerland have preserved their rate of discount unchanged for over a year—from a sheer sense of inability to understand its workings; it is wise not to meddle with forces which one has not learned to master.

(3) Keep the State at arm's length from the springs of the currency; no co-operation between the State and the Bank of Issue.

Thus my scheme ought to conciliate the spirit of our age. It is an essentially, a fundamentally conservative scheme. It makes no pretence of delivering society from ancient "wrongs" such as interest. Its political drift, if it has any at all, is away from State control and towards individual responsibility. However, radical conservatism is a kind of radicalism, is it not? And here I wish to point to the truly revolutionary side of my plan. It not only tolerates gold, it insists on gold as an essential factor in the mechanism of a stable currency—at a time when even conservative economists have lost their faith in gold, and no one dares, or cares, to plead for its retention except in an apologetic strain. It not only preaches no sort of enmity against interest, but it establishes interest as the very soul of money, not to be tampered with, much less oppressed, at a time when, in consideration of the enormity of the debts piled up by the war, even conservative economists and statesmen have begun to oppose the claims of interest. Indeed, this scheme of mine is so reactionary as to glorify the two things which all the advanced spirits of our enlightened age are unanimous in reviling : gold and interest, the arch-exploiters of humanity.

In one respect, the purely theoretical one, my scheme is altogether subversive and runs counter to all the accepted notions of all the accepted interpreters. Instead of allowing such a reasonable and obvious idea as that "dear money," i.e. a high rate of interest, makes for a restriction of the output of money, it bases its practical provisions on the observation that a thing which is made dearer is all the more eagerly produced, and that borrowing becomes the less attractive the more the lending rate is reduced—in exactly the same way as buying is rendered risky when prices decline. Such heresies as these are, I hope, sufficient to mark my attempt as unworthy of the notice of the professors. My solution of the problem is of the same nature—*si parva licet*—as the suggestion of those who first conceived the earth as revolving round the sun instead of the sun round the earth. The professors and the publicists of that time knew better than the Galileos ; but history has decided who was right.

E. D.

BASLE, SWITZERLAND,
February 1925.

CONTENTS

PART IV

POSTSCRIPT

PART I

PRELIMINARY

I

THE HERETIC AND THE PROFESSORS

In order to make my way of reasoning and my demonstration understood, I am bound to preface this treatise with a few remarks on Silvio Gesell, the first and greatest of those economists who have been advocating the stabilization of the currencies. He began as far back as 1891. By 1898 his theory of a stable currency was fully developed, as may be seen from his pamphlet *The Monetary Problem in Argentina* (*La cuestión monetaria argentina*). (Gesell had been established in Argentina as a merchant for some ten years.) His most comprehensive and most attractive statement of the problem is contained in *Aktive Währungspolitik* (*An Active Currency Policy*) ("a managed currency," in Mr. Keynes's phrase), published in 1908. All the arguments to show the failure of the traditional system and the advantages of a stable standard are there set forth with masterly clearness and completeness. Unfortunately Gesell was led astray by certain hasty conclusions from his truly grand discovery, and, not satisfied with a simple monetary reform, he reared a structure of social and political reform which has more and more obscured the real issues of his scheme. I said, unfortunately. From my personal point of view I ought to say, fortunately. For it was by the social programme that I, a confirmed Socialist, was first drawn to the study of Gesell's work, while it was by his errors that I was forced to reconsider the whole problem. I owe it in no small degree to what in Gesell's theory is fallacious that I am now in a position to offer a solution which hitherto seems to have escaped the investigators of the problem. I wish it to be clearly understood that my scheme is derived from Silvio Gesell's pioneering work, to the greatness of which, as to the nobility of thought in which it is conceived, no one can pay a sincerer tribute than I do. But also I feel it necessary to state that I have rejected Gesell's land nationalization scheme, his theories of gold and of interest, and his Reparations programme, and that I have dissociated myself from the so-called Free-land Free-money movement in Germany and Switzerland.

It is not for me to tell English readers to whom honour is due in England and America for having been pioneers in the campaign for monetary progress. The names of Professor Irving Fisher and Mr. J. M. Keynes are well enough known, and not only in their respective countries. Rather different from their contributions, less complete in the details, obviously less expert in its treatment of the matter, more like Gesell's in its tone and general attitude, is Professor Frederick Soddy's little book *The Inversion of Science and a Scheme of Scientific Reform* (1924). But it is immensely suggestive through the high originality of its thought and the powerful denunciations of the abuses and evils of the old monetary theory and practice. When a man of the standing of Professor Soddy turns away from his special work to inform the world of science that an inadequate monetary system threatens to rot the fruit of scientific effort and progress, and when the searching intellect of the born discoverer of scientific facts is led to think that this system, unless it is reformed, is going to wreck the whole of civilization, many will be startled into some amount of interest in the problem who would not otherwise be roused.[1]

[1] Henry Lowenfeld, *Money in Fetters,* did not come to my notice till my book was finished. I have dealt with it in a few additional pages.

II

THE SCHEME OF AN INDEX-NUMBER STANDARD

I SHALL now sketch, very summarily, the schemes of stabilization as proposed by the authors above mentioned. Professor Irving Fisher's book on *The Purchasing Power of Money* has been before the students of monetary theory for a good many years. On its findings is based the *Bill to Stabilize the Purchasing Power of Money*, introduced in the American House of Representatives, May 25, 1922, by Mr. Goldsborough of Maryland, which will be fully discussed below. Professor Keynes has been converted to the Index-number standard more recently—at the time of the Genoa Conference he did not yet declare for it—but his book *A Tract on Monetary Reform* (1923) bids fair to produce an effect much more far-reaching than any previous publication. In all essentials his scheme corresponds to that of Silvio Gesell as set forth in *Aktive Währungspolitik*—in Gesell's most comprehensive work, *Die Natürliche Wirtschaftsordnung durch Freiland Freigeld* (*The Natural Economic Order through Free-land Free-money*), there is a great deal of matter that lies outside the problem of currency. The following are the points on which all agree:

(1) The main requirement of a standard of currency is the stability of the purchasing power of money; in other words of the general level of prices, as it is to be ascertained by the computation of Index-numbers.

(2) The stability of the internal price level must not be sacrificed to the stability of the rates of exchange.

(3) A currency based on gold cannot be stable, because the price (value) of gold is subject to uncontrollable fluctuations; therefore the standard must be freed from the exclusive influence of gold.

(4) To insure an unbiased and impartial management of the currency a close collaboration of the National Treasury and the Central Bank of Issue is essential.

The Index-number Currency may be characterized as a strictly nationalized and State-controlled system, managed on certain principles, with gold excluded as a circulating medium; these outstanding characteristics distinguish it from the Gold Standard Currency, which is supposed to be automatic and free of State control, and from my scheme of a truly automatic and perfectly uncontrolled Interest Standard Currency, as set forth in this book.

Of the Index-number standard more is to be said below. Very succinctly stated, this is the theory of its mechanism: Price is the outcome of the supply of money. It rises when the supply grows, and falls with the supply. Therefore the rise of prices can be checked by the reduction of the supply of money, while the fall of prices is counteracted by increasing the supply of money. The trend of the price movement is regularly ascertained by the statistical computation of the average price of a certain number of staple articles of consumption. According to the findings of the Bureau of Statistics the controlling authority will issue or withdraw money, whenever the Index-number shows a tendency one way or the other.

PART II

THE UNSOLVED PROBLEM OF GOLD

I

THE DISPOSAL OF THE GOLD RESERVES

THE monetary systems of the past and present time have been based on a metallic foundation; the new system, rightly understood and carried to its logical perfection, has no use for gold or silver. Of the four schemes which we are here considering, Silvio Gesell's is by far the most thorough-going and radical; it leaves no loophole for gold to come in by. Professor Fisher and Mr. Keynes endeavour to spare the old monarch; however, an examination of the plea which they put up for the retention of gold will show that the case is not proved.

From the repudiation of gold as an essential ingredient of the currency arises the necessity to dispose of the precious metal. This brings us up against a difficulty, which it seems to me the advocates of the Index-standard do not fully realize. That the State should be given absolute control of the money system is a drawback only in the eyes of those who do not love the State. But that no one knows how to deal with the old money machine, that is the gold reserves, is one of the hitches on which the money reform idea comes to grief.

All down the ages the money problem has been a gold problem. What has stood the test of time as gold has done cannot be easy to uproot and thrust aside. Silvio Gesell, the Radical, who has little reverence for old institutions, proposes to sever at one blow the connection between money and gold. By his scheme gold is entirely eliminated from the monetary system, and ceases to be money. All its traditional privileges are to be cancelled; gold is reduced to the status of a common ware, such as straw, blubber, dynamite, motor-cars, newspapers, lettuce, false teeth, and treatises on the theory of value. The State will not admit gold as legal tender, so that gold cannot be used to pay taxes with. The State knows that only its own money can hold its own, so the State need not trouble to defend its notes against the competition of demonetized gold: so Gesell argues. This is overlooking the special properties of the gold material, and the mistake is all the more astonishing in the man who traces all the principal evils of the metallic currency to the natural and intrinsic superiority of the precious metal over other commodities. To imagine that this superiority will vanish, as it were to oblige the State, merely because the State abandons gold, is a strange delusion. Gold will be gold and go to market as the ducks to water. Whether coined or uncoined, whether in lumps or wrought into objects of art, profane or sacred: gold will turn itself into money, to buy, when occasion serves.

Moreover, the gold is with us. It cannot be turned out of our sinful world any more than out of the thoughts of men—and women. It cost rivers of sweat and not a little blood to obtain it. In vast piles it is heaped in the vaults of the Banks of Issue; in numberless little hoards it lies in ambush in the homes of thrifty savers. What is to become of all this treasure? Gesell passes over the problem with a jest: all brides are to be presented with some pretty gold trinket on their wedding day. "If the State sells the gold to the highest bidder," says he, "the price of gold will be depressed, which will embarrass other nations. . . ." Would it not seem that gold which is given away should depress the price rather more than gold that is sold at a price? Brides will become mothers and find it convenient to part with their finery for some more necessary piece of goods. By whatever outlet the gold of the Reserve issues into the open, it will depreciate the gold already out. Thus Gesell's expedient is no solution. It leaves the corpse of the murdered gold money lying in the path of the money circulation. A ghost will rise from it and create strange and unholy horrors, which no State authority will have power to allay. Gold has never yet obeyed the decrees of monarchs or parliaments ; it will not, it need not, be awed into submission by threats of demonetization.

Mr. Keynes's plan is hardly more reassuring. We are warned off by a certain ambiguity. Very cautiously and wisely Mr. Keynes bases his proposals on the lesson taught by successful experiment :

> I believe that in Great Britain the ideal system can be most easily reached by an adaptation of the actual system which has grown up, half-haphazard, since the war.

The position of gold is described in these terms :

> The Bank of England's gold is immobilized. It neither buys nor sells. The gold plays no part in our system.

With this compare his proposal with regard to the existing reserve :

> If we agree that gold is not to be employed in the circulation, and that it is better to employ some other criterion than the ratio of gold reserves to note issue in deciding to raise or lower the bank rate, it follows that the only employment for gold (nevertheless important) is as a store of value to be held as a war-chest against emergencies and as a means of rapidly correcting the influence of a temporarily adverse balance of international payments, and thus maintaining a day-to-day stability of the sterling-dollar exchange. It is desirable, therefore, that the whole of the reserve should be under the control of the authority responsible for this, which, under the above proposals, is the Bank of England. . . . Nor would the amount of gold, which it would be prudent to hold as a reserve against international emergencies and temporary indebtedness,

bear any logical or calculable relation to the volume of paper money: for the two have no close or necessary connection with one another. Therefore I make the proposal—which may seem, but should not be, shocking—of separating entirely the gold reserve from the note issue. . . . The reader will observe that I retain for gold an important rôle in our system. As an ultimate safeguard and as a reserve for sudden requirements, no superior medium is yet available.

This, surely, is not quite consistent enough. To-day the "gold plays no part in the system," because it "neither buys nor sells"; in Mr. Keynes's own system "gold is not to be employed in the circulation." Where is the difference? The gold at the Bank will no more buy than it does to-day. Nor does the reason given for keeping it at the Bank strike me as valid. With a currency which safeguards economic life from all risks, it is hard to see how the need of "a war-chest against emergencies" should ever have to be faced. In the case of a war, the gold would be held all the more tightly. Where, then, is the reserve gold ever to come into play? If, moreover, the other States follow suit in adopting a stable standard, so that rates of exchange remain constant, and no country can ever have an interest in taking gold in payment from other countries: the reserve will never need to be approached at all. Thus the "important rôle retained for gold" in Mr. Keynes's system seems to me to be that of the buried corpse. Is Mr. Keynes luring the good public? "An important rôle for gold?" Why, gold is the rock of safety: let us have, by all means, the stable currency standard that is backed by gold! Sooner or later a real solution will have to be found. For we may count on the gold corpse, however deeply buried, emitting a ghost. Gold is a peculiar stuff, which cannot be stifled and must not be trifled with. Better not create ghosts, but have it downright and plain from the beginning.

Professor Soddy proposes to expel gold entirely from the national currency, but like Mr. Keynes he thinks that it might be usefully employed in settling international relations. He therefore suggests that the gold reserves of the world should be pooled and placed under the guardianship of the League of Nations. A national stock should, however, be kept so that a nation may be able to "buy and sell its own money freely abroad for gold." That he has a proper notion of the gravity of the problem, Professor Soddy expresses in this sentence: "America, left after the war with most of the world's gold, would probably go to war if it were left on her hands by being demonetized in other countries."

It is only when we come to Mr. Goldsborough's Bill that we are duly impressed by the witchery of gold; for here it is fully displayed. The main proposition of the Bill is to this effect:

> The gold dollar of the United States shall cease to be a constant quantity of gold of variable purchasing power, and thereafter shall be a variable quantity of standard gold bullion of approximately constant computed purchasing power.

These are the provisions regarding the disposal of the reserve gold:

> That the Secretary of the Treasury shall divide all the gold against which gold coin certificates and gold bullion dollar certificates are outstanding . . . into two parts, one part to be known as the reserve against outstanding gold bullion dollar certificates and equal to 50 per centum of the value of the gold certificates then outstanding, and the remaining part to be known as the surplus in excess of said reserve. This remainder or surplus shall forthwith be transferred to the general fund of the Treasury as the initial profits of the new system.

It must be very cheerful to begin with profits. The last provision can only mean that the State is to employ this surplus gold for its own needs, that is spend it. Thus half the gold treasure would be circulated, though not as legal tender. Can this be done? Is it not a deception? We shall show, in a later section, that it is. Like Gesell and all the others, Mr. Goldsborough assumes that on gold being demonetized, the price of gold will be entirely detached from the currency and vary freely. (In another sense, however, he reasons as if nothing would happen to the value of gold, when the reserve is torn in two halves and one half offered for sale.) The gold contents of the dollar shall from time to time be officially refixed by the computation and use of Index-numbers, so that the goods contents of the dollar, i.e. the purchasing power of money, need not fluctuate. This amounts to saying that the price of gold will vary, in relation to the average of other prices, as it has always varied—or seemed to vary.

The gold weight of the dollar will grow and diminish according to circumstances. A "variable quantity" of gold is to be obtained for a goods dollar (Index dollar, currency dollar, legal tender dollar, or compensated dollar). When gold becomes dearer the weight of the gold dollar is reduced, so that less gold is obtained for a goods dollar —I have to be explicit to be sure that no point in the argument escapes the reader. Naturally so; for the growing dearer is expressed in the very fact that less gold needs to be surrendered to obtain an equal quantity of other things. The official fixing of the quantity of gold contained in the gold dollar is nothing but a confirmation of the fact that the price of gold has risen.

But what is to be the use of this gold dollar? There is another idea, perfectly extraneous to the purchasing power of money, lurking in the scheme. The notes issued are to be backed—though only up to half their real amount—by a reserve: they are to be redeemable, much as if they were a venial sin. No reason, however, is alleged to justify this "reserve against outstanding gold bullion dollar certificates"; "reserve against": as if the certificates were an advancing enemy! It is only provided that

The reserve shall be maintained daily as nearly as possible at 50 per centum of the gold bullion dollar certificates outstanding from time to time.

What can be the object of this provision? In order to find some clue we ought to set it off against the provision that the contents of the gold dollar are to be variable, and from time to time readjusted to the currency dollar. There are, therefore, two standards prescribed for one and the same object: the reserve is to be composed of gold dollars of exactly the same value as the goods dollars; the reserve is to contain exactly half as many gold dollars as there are goods dollars in circulation. This feat can only be executed by means of some juggling trick. This is it: the Bureau of the Mint, in order to readjust the reserve to the altering state of things, does not need to alter the contents of the reserve at all; it will suffice to adapt the weight of the gold dollar to the altered situation. It reduces or enlarges the lump of gold that shall be a gold dollar, as the case may require. A suitable computation can at any time decide that a given quantity of gold shall be equivalent to any number of currency units.[1]

Practically the case is of course not quite so simple. Gold dollars are made of the gold outside the reserve, and these gold dollars will be dollars, though they cannot appear in the glory of stamped coins. They will buy and so compete with the approved dollars, the paper or goods dollars. And what will happen then? Something has to be done; besides computing the weight of the gold dollar, the price of gold would seem to require some managing. Mr. Keynes in fact suspects that the United States is already practising this method. He writes:

For the past two years the United States has pretended to maintain a gold standard. In fact it has established a dollar standard;

[1] This point is more fully discussed in Section 2 of the Postscript. Of course gold is too solid a stuff to submit to the cunning shifts of a theory.

and, instead of ensuring that the value of the dollar shall conform to that of gold, it makes provision, at great expense, that the value of gold shall conform to that of the dollar. . . . No doubt it is worth the expense—for those that can afford it. The cost of the fiction to the United States is not more than £100,000,000 per annum.

What is aimed at is, therefore, to establish and preserve the parity between the gold dollar and the paper dollar. To gain the end it appears necessary to manage the price of gold. How that may be done is practically demonstrated by the present U.S.A. gold policy: gold has to be bought and kept, or sold and circulated, by the Mints. However, Mr. Goldsborough proposes a different method:

If on any date the reserve falls short of 50 per centum it is to be restored by withdrawing from circulation and cancelling gold bullion dollar certificates.

These certificates, I have come to conclude after a great deal of puzzlement, are not supposed to be money at all; i.e. they are supposed to have no influence on the purchasing power of the paper dollar. Yet, the Bill speaks of them as circulating. Circulating, on any conditions whatever, they cannot but have the same effect on prices as any other kind of means of exchange. Therefore the withdrawal and issue of these certificates, merely with a view to readjusting the value of the reserve, is a flat repudiation of the principles of Index-number Standard. The manipulation in question, the withdrawing of certificates, of "purchasing power," must influence the level of prices, which it is the object of other manipulations to maintain. We are in a vicious circle, from which we shall escape only by the assistance of a saner and simpler conception of the rôle to be assigned to gold. I might here dismiss the case by repeating and endorsing the opinion of Mr. Keynes that "the amount of gold would bear no logical or calculable relation to the volume of paper money; for the two have no close or necessary connection with one another."

What can have suggested to the author of the Bill the quaint device of those gold bullion dollar certificates? An atavism: the obsession of traditional gold standard notions. It may, moreover, serve the useful purpose of reassuring the good public: if you are going to tamper with anything, don't tamper with our gold! To save the reserve, sacrifice the notes and throw stability to the dogs. The real and approved gold standard style! Not through the acquisition of gold is the reserve to be replenished, which would surely be the natural proceeding. Withdrawing from circulation and cancelling gold bullion dollar certificates must necessitate the further diminution of the reserve. These certificates have been issued in exchange for gold:

> The Bureau of the Mint shall receive . . . all gold bullion offered to it, and shall pay for the same with gold bullion dollar certificates.

Thus the Mint, in order to withdraw certificates from circulation, is forced to yield up gold out of the reserve. A curious way to replenish it. Besides lessening the amount of gold in it, it lessens the price of gold. For the more gold is allowed to appear in the market, the less must its price be, and so the sale of reserve gold for certificates will cheapen gold. This cheapening reduces the value of the gold in the reserve, after the sale of gold has reduced the quantity.[1] The right way to proceed would be to add gold to the reserve. The purchase of gold by the Mint would have a twofold effect: (1) it strengthens the demand for gold and so raises its price, whereby the value of the reserve is enhanced; (2) it augments the quantity of gold in the reserve.

To judge the situation rightly we have to make out what is wrong and should be righted. Mr. Goldsborough has muddled the issues, because he pursues two aims which are opposed to each other. The main point is that the

[1] The transaction suggested is so preposterous that it is utterly impossible to carry it out.

level of prices shall be maintained by a correctly dosed circulation of cash. Mr. Goldsborough wants to reduce the dose, as if there were too much money out. He reasons in terms of the gold standard. The reserve used to become insufficient when the amount of bank-notes issued exceeded a certain limit prescribed, not by considerations of price, but by a convention. Under the new regulation the case must assume a different aspect. By my reading of the case the shortage of the gold reserve is a sign that there is too little money (notes) in circulation; by Mr. Goldsborough's interpretation it would indicate an excess of bank-notes. Who is right?

Let us observe the facts of reality. The question to elucidate is: does a shortage of the gold reserve coincide with an excess or with a shortage of circulation? It is generally admitted that there was an excess of circulation in Germany in the years 1919–23. Now the fact is that as the note issue increased the gold reserve seemed to improve its ratio. Although no gold was added, the percentage rose steadily as the volume of paper money was swelled. There was a time when the gold of the Reichsbank was many times as much as the total of the notes in circulation—at which the experts were greatly and very ridiculously nonplussed. Here is a case to prove, beyond the shadow of a doubt, that an excess of notes is accompanied by an excessive reserve. Hence vice versa: an insufficient reserve—provided that a definite percentage has been decreed by law—is an indication of a shortage of the circulation. It appears that the tables are turned when gold is not bought at an unalterable and fixed price. What holds good with a gold standard currency does not apply when gold is not the ruler.

The difference between the gold standard and Mr. Goldsborough's scheme consists in this. By the gold standard the parity between gold and any other legal tender was all that mattered; it was on no account to be broken. By the American Bill the ratio of gold to paper is supposed to vary, so that the general purchasing power is not to be affected by movements of the price of gold. Consequently it may happen that a given amount of notes will buy a smaller quantity of gold than at an earlier period, or vice versa. Contrary to what we observed in Germany, it would be the purchasing power of gold that varies, while that of paper money continues stable; but with regard to the adaptation of the ratio between the two, the same principle would obtain: the gold reserve might rise above or fall below the percentage merely through the rise or fall of the price of gold. This amounts to a complete reversal of the case: an excessive circulation is associated with an excessive reserve. An excessive reserve always means two things: a large reserve as to the quantity of gold it contains, and a high price of gold. These two must be coupled because the price of gold advances when gold is drawn to the reserve (i.e. withdrawn from the supply), whereas the sale of gold, which depletes the reserve, reduces the price of gold by increasing its supply.

Obviously, what was observed in Germany (and elsewhere) is logical and inevitable. Few notes—good notes; the less, the better. Better here signifies: more substantial, endowed with greater energy and purchasing power, even as compared with gold. The gold reserve becomes insufficient because the notes issued are over-sufficient, i.e. too substantial. But this excess must not be assailed by the withdrawal of notes; that would be supplying the enemy with ammunition. Withdrawing notes strengthens the notes and increases their "weight." They must be weakened, and that can be achieved only by swelling their numbers.

This will be more easily understood if the reader considers that it is the value of individual dollars that has to be levelled: every gold dollar of the reserve is always to have the same value as every paper dollar in circulation.

If the reserve is computed to be short, every gold dollar in it must be insufficient. The reserve can be brought up to the mark only by improving the value of the individual dollar. By reducing the number of goods dollars, according to the provision in question, it is the goods dollar that is strengthened, while the gold dollar is further weakened. The price of gold is determined by the supply of and demand for gold in the market. In order to restore the value of the gold in the reserve, the supply of gold must be lessened, which is achieved by gold being drawn to the reserve. This would have suggested itself to the author of the Bill, if he had not been misled by the practice of the gold standard system.

I confess to an utter incapacity to understand Mr. Goldsborough's Bill.[1] Instead of wearying the reader with a prolonged examination I will briefly review Professor Fisher's article on "Stabilizing the Dollar," published in *Harper's Magazine*, March 1924. Professor Fisher seems to endorse the Goldsborough Bill; he does not take exception to any of its provisions. Is it because he considers it consistent and workable? I do not know, but I am glad to see that on one or two points he suggests a solution differing from the provisions of the Bill. Whereas the latter's reserve is to contain only half as many gold dollars as there are goods dollars, Professor Fisher intimates that the ratio is to be 1 : 1. I must quote the passage :

> Vary the weight of a coin ? Now, of course, we cannot vary the weight of a *circulating* coin. But few gold dollars take the trouble to circulate, and none need to. Most of them are content to repose comfortably in the Treasury, and all might well repose there. For it is by thus reposing that they perform their chief function, which is to confer upon the paper dollars which do circulate whatever purchasing power these have ; that is, whatever purchasing power the gold dollars would themselves exercise if they circulated. The purchasing power of paper is vicarious. It exists because any

[1] Another attempt to fathom its depths is made below.

paper dollar can at a moment's notice be redeemed at the Treasury for one of these restful gold dollars. . . . Now the law which entitles you to a gold coin in redemption of your paper dollar may be changed so as to entitle you not to a gold coin reposing in the Treasury, but to a quantity of gold bullion reposing in the Treasury—not a fixed quantity of gold bullion, but whatever quantity happens, at the moment of redemption, to be equal in value to one goods dollar. What this quantity would be at the moment would have been previously ascertained by the use of the Index-number.

Convertibility and redemption : we are back in the very heart and fortress of the real gold standard. If Professor Fisher means to say that every goods dollar is redeemable, we are forced to inquire how the issue of currency is to be effected. I fail to comprehend it. Goods dollars, I take it, can be issued irrespective of the amount of gold in the reserve ; for surely the gold in the Mints will not grow bigger when crops are plentiful and demand an increase of circulation. But then it sounds like a mockery to say that every dollar is redeemable, a mockery to drag in gold at all. At that rate the reserve might be allowed to dwindle to a mere nothing ; by Professor Fisher's definition a pound of gold might be made equal to all the marketable wealth of the country. For to say that "whatever purchasing power the paper dollars have is conferred upon them by the reposing gold dollars," amounts to an affirmation that the gold in the reserve is at any time worth as much as all the circulating goods dollars combined will buy at any one moment. (They would buy mighty little, if all were offered at once.) However, so long as it is not provided—as the Goldsborough Bill does provide—that any currency bills shall be made or unmade to suit the requirements of the gold reserve, all this computing of the weight of imaginary gold dollars is a harmless sport. Practically the scheme is the same as Mr. Keynes's : the gold reserve is entirely separated from the note issue, and the amount of gold to be held in the reserve does not "bear any logical or calculable relation to the volume of paper money." One of the chapters in Gesell's *Aktive Währungspolitik* (1908), is headed : "The amount of gold of the Banks of Issue is no measure for the issue of Notes." So these three theorists are fairly well agreed.

Goods dollars are to be so contrived as to buy a constant quantity of commodities. Their purchasing power is entirely based on the existence and the quantity of purchasable goods. Take part or the whole of the goods away, and just as much purchasing power will go. The purchasing power of money resides, not in the reposing gold, which very few people have any use for at all, but in the goods and services offered for exchange. To make out the power of a horse we have to attach it to some load. It can only display as much power as the load amounts to. In the same way the power of money can only turn over as many goods as it is attached to, as are in existence. But the amount of money issued is always powerful enough to move the volume of goods in the market. When it happens that the goods do not get turned over properly the fault is not with money, but with another factor.

With the gold standard the idea of redemption had a certain air of reasonableness about it ; and the regulations were hard and fast. It made people believe that they possessed a definite value when they possessed a certain quantity of gold. Now Professor Fisher wants to make people believe that they possess a definite value when they know the number of gold grains that a dollar contains. Practically it amounts to the same : it is the same uncertainty, instability over again. The confusion produced in the heads of the muddled and hasty public by the notion that there are two dollars, which are to be the same though one shall vary, will be a painful spectacle. Professor Fisher's solution may be merely a concession to old prejudice, and intended as a sop. But I should deprecate making use of a subterfuge ; it is time people should have a chance to learn the real facts concerning money : that it is the wares and conveniences which they depend on for their daily use, and which they supply by dint of much application to hard tasks, that create the purchasing power of money ; that if they wish to have good and honest money they must produce good and honest things for use, and use them so sparingly as never to let consumption outstrip production. Where does gold come in ? When I desire to possess gold, well, let me try how I may obtain it ; but spare me from the enforced idea that when I use my paper money I am using a mere ghost, an unreality, a fiction.

Professor Fisher's conception of paper money as merely representing gold money reposing in the reserve can be confuted by this consideration. When a country's natural wealth has been impaired, its gold ceases to buy. It becomes almost impossible to procure gold. People will not part with their precious treasure while there is so little to be got for it. On the other hand, when the natural wealth of a country grows, gold comes out of its reserve, and it is easy to obtain gold. People pass it on freely because there are plenty of goods to exchange it for. If, therefore, we are told that the value of the paper dollars is borrowed from the gold dollars in the reserve, it ought to be added that the value of the gold dollars is borrowed from the wares in the market. It follows that the source of all money value, gold and paper alike, is the volume of marketable goods. If all the paper dollars of the United States were spirited away at one blow there would be no appreciable diminution of wealth. Likewise if all the gold dollars in the reserve disappeared the purchasing power of the paper dollars outstanding need not suffer, so long as the volume of marketable goods is not diminished.

Professor Fisher reinstates the terms "redeemable" and "redemption." I had thought that money reformers had got beyond this stage. Does he mean to say that the Federal Reserve Board would redeem, piece by piece, the paper dollars outstanding ? Surely no ; but the way he puts the matter is something of a make-believe. He knows that no one except dealers in gold will ever attempt to get gold out of the reserve. When there are plenty of goods in the market paper dollars are received confidently ; when goods come to fail people will, surely, not go out of their way to procure gold, though they will spare no trouble in attempting to procure the needed wares. So the chance for redemption will never come. If it did come we may be quite sure that an emergency law would immediately stop the run on the reserve. Thus the redemption clause cannot become operative. If it simply means that anyone needing gold may obtain it at the Federal Banks the word " redeem " should not be used ; such a transaction is a mere purchase of gold. The object of the buyer is not to be rid of the notes, but to procure gold for some definite purpose, speculative or otherwise. It would contribute to simplify the theory of money and currency if the simpler terms " buying " and " selling " were employed in the place of " redeeming " (or converting) and " issuing " ; also if we said " price " where the word " value " is used.

The most uncompromising abolisher of gold among the stabilizers is probably Mr. Henry Lowenfeld in his book *Money in Fetters* (London, John Murray, 1924). He has a good many quaint remarks about the dangers of the precious metal. His chapter on " The Glamour of Gold " is evidently intended to rouse the reader's indignation against it :

> But " Finance " is strongly interested in the dealings with gold. It watches with the utmost care over its vested interests, and consequently gold has been removed from its lofty height only where " Finance " was best served thereby, and in other places its influence is energetically defended. . . .

When an author resorts to this method for proving his point we may be sure that he lacks good arguments, and is at a loss for a real expedient. The sentiment expressed in our quotation proves nothing against either gold or high finance. Finance is equally strongly interested—and why, indeed, should it not ?—in all conceivable kinds of articles as it is in gold. Evidently Mr. Lowenfeld has not fully mastered the problem of the traditional gold standard. That is why he has no suggestion to make concerning the disposal of the gold—he simply advises to abandon it and leave gold to its fate—neither does he seem to realize that his proposal creates a problem. I would venture to say that the power of the people directly or indirectly interested in the retention of gold is sufficient to doom to failure any attempts to cast it out of the currency systems. This consideration in itself ought to rule out any schemes which tend in this direction. They are not worthy of a serious discussion.

II

THE PRICE OF GOLD

PROFESSOR FISHER endeavours to solve the problem as to how gold might be made to serve the currency without at the same time undoing it, as he believes it to have done in the past. He seems to reason something like this: Gold cannot be eliminated out of the system by a decree the way some hasty critics imagine and affirm that it can; gold will always be thought of and treated as money. But the price (called value) of gold has always varied, and must vary for ever. How, then, is this unstable gold to be combined with a standard of value that shall be stable? At the bottom of the problem, with Professor Fisher, lies a vague feeling that there must be some sort of natural innate stability connected with gold; otherwise the device of a gold dollar, part stable, part unstable, could not have recommended itself to him. As it is to be kept at a parity with the goods dollar, the gold dollar will have unvarying purchasing power like the former; for it is provided in his scheme that the weight of the gold dollar shall be recomputed when it has appreciably detached itself from the standard. But as the price of gold is expected to vary, the weight of the gold dollar would be variable. There is as good a Sphinx as any that has been set up: variable and invariable in one. We shall come to realize yet how very near Professor Fisher's solution comes to the true one, in spite of its roundabout method.

Also Mr. Keynes takes it for granted that the price of

gold will shift about. "Our object must be to secure this advantage, if we can, without committing ourselves to follow big movements in the value of gold itself." If it were not for the foreign exchanges he would disregard gold altogether; but the usefulness of it in levelling prices internationally induces him to seek for some way to secure its services. Alas! it is to be by the Bank "regulating the price of gold, just as it already regulates the rate of discount." Let me quote some more (p. 190):

> Regulate, but not "peg." The Bank of England should have a buying and a selling price for gold, just as it did before the war, and this price might remain unchanged for considerable periods, just as bank-rate does. But it would not be fixed or "pegged" once and for all, any more than bank-rate is fixed.

Although there is no attempt made to create a gold sterling and bring about some sort of parity, we here find the same endeavour as in Professor Fisher's scheme: to fix the price of gold in accordance with something else. I have no doubt it can be done by a powerful State like Great Britain. However, it would cost a great deal of trouble, and besides spoil the chances of gold. The best, the only true service gold can give is only to be obtained of unregulated, free and sovereign gold.

I believe that the strange intricacies of the actual situation have scared money reformers into seeing difficulties where there are none. Although practically abandoned, the gold standard rules and laws continue to sway men's minds. Old notions, but dislocated and shattered, and curiously blended with new theories, haunt their arguments. One real and embarrassing difficulty imposes itself in the fact that in no country has there been for a considerable time a steady ratio between the price of gold and the level of prices. Nobody can tell what a pound sterling, a dollar, a franc, a mark in terms of gold is. Statistics will make out the price of gold, as measured by Index-number, to have fallen in 1921, to have risen in 1923. It seems impossible, and it is impossible by the methods hitherto employed, to arrive at a satisfactory solution.

The first step to take is to set gold free. Our examination of the case will leave no doubt, I trust, as to its being not only feasible but eminently beneficial. All that is needed to avert the consequences which everybody shrinks from facing is to peg the rate of discount (see below). Within a year or two of the inauguration of such policy it would be proved to the eyes and purses of everybody that the price of gold has settled down into a perfectly stable ratio with the paper currency. By the traditional gold standard the price of gold was "pegged," while the rate of discount was allowed to vary. This method worked but poorly. Mr. Keynes believes that the remedy might be to vary the price of gold along with the rate of discount. I ask: is not this to create a wider scope for fluctuations? I will not defer my answer, but right here express my opinion that, provided the level of prices can be stabilized by the method proposed, both the price of gold and the rate of discount will thereby become stabilized too. Up to the end of the reign of the gold standard all the three factors in the equation: the price of gold, the level of prices—which two should really be considered as one—and the rate of discount moved together, under the pressure of one and the same force. They are so firmly linked, so closely related, that what moves one necessarily dislocates the others. If, therefore, we succeed in pegging down one, this one must keep the others, which are tied to the same tether, pegged also. It was believed that the value of gold *was* pegged. But no, it was fixed to a weight, i.e. a mere convention, not a reality. That the price of gold was not stable is proved by the fact that neither the level of prices nor the rate of discount was.

One of the three factors has to be reduced to stability if we are to obtain a stable currency. But only one must be "managed," while the other two must be allowed to move into the line and find their own level spontaneously, naturally, i.e., as a natural *necessity*. The attempt to manage gold has failed; the idea of managing the level of prices through Index-number is going to prove impracticable; the true solution will be found to be the pegging of the rate of discount. In order to prepare the ground for a discussion of the problem, I shall now try and show why the price of gold, if gold is but allowed full liberty to follow its bent, must naturally remain at a parity with the legal tender money.

When I say "if allowed full liberty to follow its bent" I mean that gold must not be arbitrarily deprived of its right to serve as money and legal tender, and not be managed in any way whatever. Silvio Gesell abstains from managing it—at least after it has been disposed of to the happy brides—but he oppresses it out of the currency. Professor Fisher and Mr. Keynes would rather not have it circulate; to prevent its doing so they set out to manage it. As long as it is kept under a pressure gold cannot display its virtues, but will behave like a defiant and rebellious spirit; the best are ever the worst under ill-treatment.

Gold can never be severed from the currency long. For men will never cease to deal in gold and use it as a means of exchange. It imposes itself as money by its intrinsic properties. The State may, to please a theory, demonetize gold. At the present time gold *is* demonetized, because States adhere to a theory which purports to instal it as the very essence of money. When demonetized by a decree, gold will automatically constitute itself a means of exchange, will coin itself, by virtue of its uselessness, which compels it to preserve itself from depreciation by rendering services. The gold trinkets, in being offered for sale because they are becoming common and cheap, will call forth new money: they literally convert themselves into money. Gold is so constituted that it can at any time turn to money. (Perhaps I ought to say: Men are so constituted that gold can turn itself into money at any moment.) Attempts to eliminate it from the currency will never succeed; for always gold will seek its natural employment, whenever other activities and uses, from which its value might be maintained, are denied to it. For gold, whatever scorn Gesell may pour on the notion, is a store of value. The gold in existence was acquired at a price, it has value, and loss of this value means a loss to all owners of gold. The owners of gold will remember its uses so soon as gold is threatened with extinction, even though they may have abused it for a lifetime. This close and inseparable connection of gold with money has for a natural consequence that, in the countries where gold is suffered to move freely, the price of gold must, with a stable currency, remain stable. Up to now the fluctuations of the price of gold made the fluctuations of the currency: that is how the critics of the gold standard put it. But it is at least as correct to say that the fluctuations of the currency produced the fluctuations of the price of gold (see Appendix, p. 62).

To demonstrate the truth of these propositions I shall examine the assumption that the price of gold will fluctuate while the stability of the currency is maintained by means of paper money and credit control.

A. Gold Tends to Become Dearer

This tendency is expressed by the fact that a given quantity of gold buys more goods, and therefore a greater amount of paper money, than before. The original parity between gold and bank-notes is broken in favour of the former. What will happen? Gresham's law begins to operate, by which the poorer money turns the better money out of circulation: in our case paper money dethrones gold. A thing that becomes dearer ceases to be supplied;

most of all when the material is as easy to store as gold is. The "uncheap" gold will no longer go to market, but will stay at home in its safe corner. Now what is it that is here withdrawn from action, money or merchandise? Under the gold standard the appreciation and consequent hoarding of gold would bring about—and in its turn be encouraged by—the withdrawal of bank-notes by the bank of issue, and the fall of prices; it spells slump. But very different is what takes place when the price of gold is not fixed to a weight. These are the reasons:

(*a*) As gold avoids the market it does not only cease to be money, it also ceases to be a ware. It is reduced to a mere hoard, and so is deprived of all its uses. If on one hand the supply of money—that is demand, which should be created by the gold money—is diminished, the supply of goods is at the same time diminished by the same amount, namely, the full amount of gold hoarded. This in itself suffices to prevent a general pressure on the price of other articles: the amount of the hoarded gold money and the amount of the hoarded gold ware are equal and reduce each other to nought.

(*b*) Where is the enhancement of the price of gold to come from, if it is not accompanied by—or expressed by—the depression of the general level of prices? There is no enhancement at all. Very soon it will be found out that gold, in trying to make itself rare, has wrought its own damage by betraying its weakness, its limitations. Weakness and limitations detract from the value of things. No sooner have the gold pieces gone into hiding than they begin to feel scared: they discover to their dismay that nobody cares and thieves do not try to break in and steal. The hoarders are taught that gold will no longer be available as a hoarding commodity, since it can so easily be replaced in the market by other money. For we have to remember that in the proportion in which gold coins are reserved and money tends to become scarcer, business, rather than procure gold at a premium, will obtain new bank-notes at the bank of issue and so rescue the price of the stocks on hand. Business has recourse to credit. (It is here assumed that the bank of issue is by law bound to discount bills at a fixed rate of discount.) This is the decisive turn: the fatal money shortage is thus averted and the deadening slump of prices is arrested at the start. (By the terms of all the schemes under discussion, the money office would issue new notes; practically the effect would be the same.) When the money hoarders discover that their speculation has miscarried they will repent. The hoards of gold that cannot be turned to any profitable use grow uneasy. There is nothing to be gained by keeping them, so they are anxious to be turned loose again. They are threatened by decay: for they yield no interest (nobody seems to take any interest in them), while all other things bear their regular 5 per cent. Thus they burst their safes to escape into the market, the refuge and field of action of whatever is harried by the threat of decay—the loss of value. Now, this new supply of gold out of private hoards, this seeking a market, cannot but cheapen gold. The price of gold, if it has risen ever so little, relapses to its old level.

To enable the reader the better to visualize the process here traced out, I will anticipate a point which is to be more fully discussed below. When gold, according to our assumption, appreciates, it is because commodity prices are rising. To counteract this development it would be necessary to reduce the supply of money. By reserving itself in the manner described the gold portion of the currency automatically brings about the reduction of the money supply. This reduction is needed, and so it would seem that an increase of the note issue would be a fault. It may not be necessary and not take place; but there is this to be considered: price movements are very apt to reach out too far; therefore a mechanism which comes into play when enough has been done is an indispensable part of the machine. The reason why, under the orthodox gold standard, gold did not behave in the manner described above is not difficult to discover. I will give it in the words of an expert practitioner. In his speech to the general meeting of the shareholders of the Midland Bank, Mr. R. McKenna said (*The Times*, January 28, 1925):

> We are all familiar with the conditions under which an ordinary trade commodity falls in value. Sellers offer more than buyers will take at the current price, and the price is reduced. But in the case of gold the process is not so simple. Sellers of gold *can always obtain the full statutory price for their commodity* in a gold standard currency, and *there must be a depreciation of the currency*, that is to say an upward movement in the price level, before there can be a reduction in the real return for gold. How this depreciation happens is worth considering. . . . When gold, whether of native or foreign production, is offered for sale to any of the Federal Reserve Banks, it will be bought at its full rate of so many grains' weight for a dollar. As the Federal Reserve Banks are central banking institutions, we remember that the effect of a purchase by any one of them is to create so much additional cash standing to the credit of the member banks. It is hardly necessary to repeat that this cash becomes the basis of additional loans, which create new deposits, or, in other words, increase the purchasing power of the public. Increased purchasing power unaccompanied by greater production leads to higher prices, and thus we complete the chain of events by which a purchase of gold is connected with a decline in value of the currency.

Under the gold standard gold cannot detach itself from the currency, but is forced to depreciate with the currency; in other words, it forces the currency to depreciate whenever, for some reason or other, it happens to depreciate itself. By the provisions of my scheme, gold is in a position to safeguard its own value. When threatened it simply detaches itself from the currency by ceasing to act as money, and in so doing it not only preserves itself, but the currency along with it, from being depreciated and damaged.

B. Gold Tends to Become Cheaper

Cheaper means: for sale. Everyone's gold is for sale, because its depreciation involves the holder in a loss. Hence gold goes to market to exchange itself for other wares. In other words: the depreciating gold has to look out how it may render itself useful and somewhat more sought after; that is the only way for it to succeed in maintaining or recovering its value, its price. It condescends again to the uses which perhaps it has been shunning. Above all, it will want to be money again. For it is for sale, i.e. it is eager to buy. Now the thing we buy with is called money: gold, all the gold, wants to be money, and nothing more natural. For the cheapening of gold is the expression of the fact that money is superior to the ware gold, is rather more than its equal. So gold must seek to be money and to participate in superiority. However the ranks of money are full, so part of the currency has to be ousted before there can be room for gold to come into play. Thus gold enters into competition with paper money. In order to triumph it only needs to be a little inferior to paper money; for the inferior money beats the superior. And gold has become inferior by our assumption. The advantage of gold consists in the fact that it can reduce its terms for interest, whereas the paper money issued by the Central Bank is tied to its fixed rate of discount.[1] Business will avoid using the discount and pay in gold, which is to be had at a more favourable price. In this manner the note issue is diminished, and the level of prices, notwithstanding the increased supply of gold, is not raised. Therefore no cheapening of gold—as a ware—takes place, because in the proportion that the supply of gold grows, gold enhances its usefulness as money, and as money it cannot fail to preserve its parity with the currency. So

[1] The procedure is more fully explained below, pp. 114 and 194.

Gresham's law decrees. Its natural and inalienable aptitude as money material preserves gold its value; on the other hand, the competition of paper money, now unchecked by any restrictions, will not allow it to be hoarded for purposes of usury.

We may be not a little surprised to discover that the price of gold, for the first time in the record of history, has become stable. As long as it was officially and arbitrarily fixed, the price of gold as measured by the price of other necessaries, fluctuated incessantly. The supposed fixed price of gold was a delusion. It amounted to the trick that a given weight of gold would exchange for a fixed sum of paper money, which in its turn would buy a variable quantity of goods. "The dollar (monetary unit) was a constant quantity of gold of variable purchasing power;" expressed in terms of an equation: *g* (gold) was *x* (goods). By the terms of the American reform scheme the positions are reversed: *g* (goods) is to be *x* (gold). With the truly stable currency standard *g* will be *g*; gold will not only be gold, but goods. While the currency fluctuates the price of gold fluctuates in exactly the same manner and measure; with a stable currency the price of gold must be stabilized too, even though, or rather because, it is left free and unregulated. This may seem strange and unnatural, but it is the only natural consequence. Gold, the material which naturally endures unimpaired, this most harmonious and equable of natural things, which is subject to no degrees of excellence: its proper place in the general price-list is at the point of indifference, where it is absolute, immutable, fixed.

If it is objected that the price of gold may be affected by the volume of the gold production, the reply holds good that production follows price, not vice versa. Like all other goods gold is produced according to its price, in the exact measure that it pays. Labour and capital will be deflected into the gold-producing industry when this industry pays better than other employments ; labour and capital will desert the gold business when other pursuits hold out better prospects. Thus the price of gold cannot fail to adjust itself automatically to the general level of prices and so to the parity with paper money. No matter what the proportion of the volume of gold to the general volume of trade (and of paper money) may be : the gold coins will circulate at par. If the demand for gold increases, gold will be withdrawn from the circulation by the public ; it is immediately replaced by fresh issues of notes called for by the public. If, on the contrary, the supply of gold is strengthened gold will flow into the cash circulation and displace bank-notes. These processes may be imagined as being carried to the extreme of one medium occupying the field to the utter exclusion of the other, when we should have either a pure paper or a pure gold circulation.

In the same way that the production of gold follows the price of the article, the consumption of it, too, must increase or diminish with the price ; but with this difference that it precedes price. What could raise the price of gold if it is not a growing demand for it ? Demand means consumption, no matter how the thing is used, whether industrially or as a circulating medium. Gesell, in *Aktive Währungspolitik* and elsewhere, repeatedly points to the danger to which a gold currency is subject through the industrial consumption of gold. Under the gold standard, when prices rise gold is cheapened, which induces people to acquire gold ornaments. As the boom progresses, more and more gold coins go to the goldsmiths' melting-pots. In this way the Bank's stock of gold is reduced, and along with it its margin for further issues of bank-notes. The result is that the boom, fed by a liberal supply of money, comes to a sudden close.

The industrial consumption of gold, however, does not only increase when gold is cheapened through the rise of prices, but more so even in consequence of the growth of the people's wealth. Gold wares are articles of luxury, and luxury is induced by affluence. If, therefore, the money reform to stabilize prices averts the slump so that the production of wealth is continued uninterrupted, the wealth of the people will grow rapidly and steadily. But at the same time the demand for gold finery will also grow, and it would suffice to consume the whole lot of gold coins if every worker and farmer were to buy some trinket for his wife or sweetheart. . . . Affluence devours the gold coins; the disappearance of the coins creates the slump, and the slump devours affluence. Hence the gold coin depends on poverty for its existence. Through the fall of prices and the concomitant unemployment the gold coin will pass out of the reach of the working classes. General affluence and a gold standard therefore exclude each other mutually and necessarily.

But ours is not a gold standard, although it allows, and even relies on, the circulation of gold coins. Now the question is whether the affluence which uninterrupted employment at high wages will bring to the mass of the people must be expected to drain the circulation of its gold and what, if it did, the consequences would be. The price of gold, we have found, cannot be affected so long as gold circulates as money. When it is affected, gold ceases to be money; at a blow all the gold that has been acting as money becomes a market ware, while the coins are replaced by new issues of paper money. Also the gold in the reserve may be sold. It will not break our standard, which is not based on or tied to gold. However there is no fear lest it should come to that. By the time that every lass and lad can afford to flaunt it in a golden sheen, many will disdain wearing gold. A thing is an article of luxury only while it is out of the reach of the many, and even the many will cease to care for gold when gold meets their eye on all their neighbours. There is this further consideration. When gold is greatly sought after the production of gold will be forced. Although gold cannot appreciate in terms of the currency, it can in terms of other wares. When people demand gold instead of books, or gramophones, or fox-terriers, or holidays, holidays and fox-terriers and gramophones and books will grow cheaper through slackening demand. The output of these seducers will diminish in favour of the output of gold till the point is reached where the lovers of luxuries will hesitate in their choice. What do you prefer, an extra week's holidays or a gold chain? If the chain, good; it will be as if you had gone out to the wilderness to dig the gold yourself, for while you work at your job a gold-digger is kept busy for you. If the holiday, good; it will be as if you had called a gold-digger away from his mine to come and take your place while you enjoy your leisure. The indulgence in golden luxuries will force on the production of gold, which has to be paid for in an increase of labour by the buyers of gold. Those who, if it were not for this indulgence, would be producers of food and clothing and housing, and luxuries of some other kind, will be gold-diggers, demanding to be fed and clothed and housed and amused by those who purchase their wares.

Gesell's argument that all the gold in the currency might be consumed industrially is only valid on the assumption that there is no other gold accessible. And not even on this assumption. It holds good only in so far as it applies to the rigid gold standard. When the price of gold has become stable through the stability of the currency, gold is safe from being misused, cheapened, as an article of luxury. It is not true that "gold coins depend on poverty for their existence."

I have been very detailed in my presentation of the case, even at the risk of wearying the reader. But have I succeeded in making it plain that the stable price of gold in a stable currency is a thing altogether different from the illusory constant price of gold in a gold standard currency? The two are as far apart as a man in the free control of his limbs and one nailed to the cross. Whence this amazing difference? Is not the case I have been making out a mere conjuring trick? I confess it took me some time to believe in my own conclusions. But note well what has changed. The gold standard demands of gold more than it is capable of achieving. The consequence is that it most fails to come up to the mark at the point where it most ought to prove its worth. In the scheme of the interest standard gold finds the ready support of paper money, and between the two they perform what is required of money. The currency constitutes, as it were, a regulating reservoir for the flow of gold. When the afflux is swelled, from whatever source, the surplus is caught up in the weir; when the afflux decreases the deficiency is made up for out of the currency, with the printing press as its inexhaustible source of supply. The outflow, which determines the efficiency of the money circulation, is always exactly equal to the requirements, because it is regulated by the natural demand for money. Anyone who has a parcel of goods to preserve from depreciation has access to the source of supply —at the price which safeguards this facility from being misused. The outflow is most accurately determined by the parity of the two kinds of money, gold and paper. And this is the manner of its regulation: more gold will flow out in the shape of money when the afflux is strong, so that the price of gold tends to be lowered; less gold will flow into the circulation of cash when it becomes scarcer.

With the interest standard gold can maintain its place in the currency, but only so long as it does not disobey the laws of nature by attempting to flow uphill, i.e. to appreciate, to lord it over its fellow, the modest paper money. A vastly different thing is gold when it holds absolute sway as a monarch, than when yoked together with one that is its equal.

I will here add that a pure paper currency would assume an attitude as tyrannous and rigid as a pure gold currency. To be adaptable and reliable a mechanism must be made up of compensating parts. Some think that in a monetary system compensation is brought about by the interplay of money and "credit." But these two, far from compensating each other, cumulate each other, each expanding and contracting with the other. When there is plenty of credit there is plenty of money; when credit fails, money also gives out. This fact has been beautifully demonstrated by Gesell—although he puts money before credit, which is a very serious error. Gesell has drawn from his observation the logical conclusion that money and credit are the same thing. As for gold, it is purely credit. When credit flinches ever so little, gold is scared. It comes and goes with credit, its very shadow. We therefore cannot rely on money and credit as compensating parts of the machine. But two intrinsically different kinds of money can compensate each other: gold, the material money, which is at the same time a consumable ware, and paper, the functional money. How these two, when properly combined, will compensate each other's failings is the subject of the following chapter. The subject of money and credit will be more fully dealt with below.

I find my contention regarding the natural parity of gold with the currnecy, i.e. perfect stability of the price of gold in a stable currency, well borne out by what John Stuart Mill says on the subject, *Principles of Political Economy*, Book III, chapters viii and ix, although his way of stating the case differs from mine so widely that few readers would be likely to detect the agreement. Mill treats money as a commodity, money by his conception being gold:

> In point of fact, money is bought and sold like other things, whenever other things are bought and sold for money. Whoever sells corn, or tallow, or cotton, buys money. Whoever buys bread, or wine, or clothes, sells money to the dealer in those articles.
>
> As the whole of the goods in the market compose the demand for money, so the whole of the money constitutes the demand for goods. The money and the goods are seeking each other for the purpose of being exchanged. They are reciprocally supply and demand to

one another. It is indifferent whether, in characterizing the phenomena, we speak of the demand and the supply of goods, or the supply and the demand of money. They are equivalent expressions.

What disguises the underlying truth of Mill's conception is that misleading and meaningless idea of cost of production. He says: "But money, no more than commodities in general, has its value definitely determined by demand and supply. The ultimate regulator of its value is Cost of Production."

I fail to discover any meaning in this, because I fail to understand how cost of production could be anything but the result of supply and demand. Say price of money instead of value of money (which is justified by Mill's own insistence that money is bought at a price), put price of production instead of cost of production (cost is merely the sum of the prices of the various commodities consumed by labour in the course of production), and the fact is manifest: "The regulator of its price is price (of production)." His own argument carries Mill to the point where he states the fact himself:

A pound weight of gold will, in the gold-producing countries, ultimately tend to exchange for as much of every other commodity, as is produced at a cost equal to its own; meaning by its own cost the cost in labour and expense, at the least productive sources of supply which the then existing demand makes it necessary to work.

"At a cost equal to its own:" what does it mean but that a pound of gold costs a pound of gold to produce? Indeed, the cost of production notion never means more than that a pound costs a pound. Mill knows and shows neatly enough why that is so:

The prices of things will, on the average, be such that money will exchange for its own cost in all other goods: and, precisely because the quantity cannot be prevented from affecting the value, the quantity itself will (by a sort of self-acting machinery) be kept at the amount consistent with the standard of prices—at the amount necessary for performing, at these prices, all the business required of it.

That is to say, price determines production, every commodity being produced in just such quantities as will pay for the cost of production. All this amounts to the simple truth that gold exchanges at par with all other commodities; for the cost of production is the same for all, when expressed in terms of money. It is the natural and inevitable parity of gold with the "standard of price," the currency. The gold standard parity is an artificial, an unnatural one—it gilds refined gold—and its existence has been the reason why the true conception of the case has been missed.

Let me try to bring out the point to still better evidence. When a shoemaker has produced a pair of shoes which he sells for a gold piece, the buyer of the shoes produces the gold piece out of his purse; it is the same as if the shoes were sold to a gold-miner, who has to fumble and dig for the bit in a rather less accessible place. All those who labour to produce commodities produce money, and they produce gold, i.e. the value of gold, if gold is employed as money. Suppose a country where gold is the only money: the producers of the metal have to be fed and supplied by those who produce the wherewithal. It is as if the whole community were a joint stock company employing men to mine the gold for them. The price of gold is determined by the cost of producing what the miners demand.

What the value of gold money really is, and how it is produced, may best be exemplified by a fiction. Suppose a country cut off from intercourse with any other inhabited region; suppose the people to use gold exclusively as their medium of exchange. And now suppose that some Prospero raises a landstorm that carries off every bit of their exchangeable commodities. Where is the value of their gold now?

It has vanished, and we see that what has been so persistently termed the intrinsic, i.e. in-dwelling, value of gold resides entirely outside gold—in the commodities that are exchanged through its agency. The islanders now set to work to reproduce what has been spirited away. In doing so they reproduce, or re-create, the value of their money. So it is proved that the value of gold, so far as it serves as money, is entirely created through the creation of exchangeable commodities. The cost of creating this value must be exactly the same as the cost of creating the wares, and the price of gold must always coincide with the standard of price.

We have now come to the point at which I may venture to challenge certain statements of Mr. Keynes's as to the price of gold. He is greatly preoccupied by the excess of gold in the United States, and he thinks that the future of the price of gold depends on the monetary policy which America will adopt. His advice to the Americans is to let the price of gold drop from its artificial level to its natural, much lower one. It will be shown in a special section below that America can get rid of her gold only by increasing her consumption beyond her production. It certainly cannot be done by merely reducing the selling price of gold, as Mr. Keynes suggests: "If gold is not wanted and must be got rid of, it would be much simpler just to reduce the dollar price of gold." If the gold is disposed of to foreign purchasers something is received in exchange: what can this be but wares demanding immediate consumption? America would be expanding the volume of her imports, and what that means to American manufacturers I need not indicate. Can Mr. Keynes be right when he opines (p. 168):

> If the United States mints were to be closed to gold, everything, except the actual price of the metal, could continue precisely as before.

To me it seems that nothing could continue as before. Prices are forces, energies, and very closely connected among themselves. The price of gold is a constituent part of every other price, so that what alters it must needs alter prices in general. A reduction of the "dollar price of gold" would be translated into a rise of the prices of wares. This may be accounted for in various ways. Gold is sold cheap; another mode of expressing the same fact is to say that gold buys dear; now gold that is sold is money which buys; to sell gold is to create new demands for goods, and whatever renders demand more urgent forces up prices. So often as the supply of gold has been augmented prices have risen; it will happen so whether gold be considered as first or last in the currency. It is a manifest illusion to imagine that the gold which the American mints are advised to sell—or merely to refuse to buy at a fixed price—would keep the peace and leave all else unaffected. Where price is affected, everything is.

But can gold be cheapened in the way here indicated? Who could be willing to purchase cheapening gold? The foreigner can buy it, but only if America will please buy of him the wares which he has for sale. The question is: Does this fit into the scheme? It does not, and so the foreigner will not do. The American citizen has no use for cheap gold. Who would surrender stable dollars for depreciating gold? For however cheap it can be bought, it will grow cheaper and cheaper in the market, as more and more gold is drawn out of the reserve. He who buys first risks to make a losing bargain; so every one will want to wait and buy last, which means that no one will make a beginning.

Late experience might have taught economists that the safe way to get rid of a ware, not excluding gold, is to make it—not cheaper, but dearer. I maintain that the present price of gold is unnaturally low, and that America can succeed in disposing of her onerous hoard only by devising a method for reinstating it in its proper station. Gold has become cheap because it has been deprived of its natural functions and usefulness, much to the disadvantage of everybody. The deplorable unsteadiness of the rates of exchange is largely due to this disabling of gold. It stands to reason that sooner or later gold will recover its position and its price; in the long run folly is overcome by practical common sense. And it is sheer folly to suppress the usefulness of things, as has been done in the case of gold. But the right way to return to sanity is not to cheapen gold, but to liberate it and so re-establish it in its uses and functions. Repeal the standard price of gold by all means; but not to fix it at a lower figure—you cannot cheapen things by merely altering a figure!—but to allow a free price to be formed. The world will be greatly surprised to find that the price will not drop.

America has, so to speak, made herself the world's dumping-ground for gold. "If you do not know what to do with your precious metal, ship it to New York; the price it will fetch is known to you beforehand," such has been the general maxim. To make a dumping-ground for a thing is to create an impression that that thing is somehow useless. I believe that with proper management of the news Press and the political platforms nations might be persuaded into dumping any one thing that you may single out. Close the dumping-ground, make it harder for people to find a placement for their gold, and gold will become dearer. So soon as there is some little uncertainty about the price, owners and would-be sellers of gold will cast about for new openings. New openings will speedily be discovered. Gold will return to its wonted and preferred occupations, and in doing so it will regain its value and esteem. It will find a wider field, it will be sought by those who have been doing without and learnt to do without it; the heaps that have been formed in a few favoured spots will gradually be dispersed.

Crowd a lot of workers into a narrow space where they cannot move freely: their ability to work and serve is gone. (I have no doubt but some ingenious person will prove from what he observes that workers are no good and should be got rid of!) The price of a ware is determined by its greatest density: the more gold becomes concentrated in a few places, the more unwieldy and useless it is bound to become, and the more will men learn to dispense with it, and the less will it be appreciated, the less will its price be. It should also be considered that gold, in order to be serviceable, must be available in sufficient frequency, i.e. it is disabled by scarcity in the same way as by excessive density. When the supply of gold falls below a certain limit, it becomes useless—like odd boots; the owners will try to get rid of it to those who hold the main lot. This accounts for the puzzling fact that gold will keep pouring into choking America from countries which have very little of it left: what little they do have left is no longer serviceable owing to its insufficiency. Replenish the European gold stores and channels to the point at which gold can flow and float again, and the demand and price of it will come up to the mark. If I had any money to venture on a speculation I should certainly speculate on gold appreciating again.

A few remarks on this following passage impose themselves:

> Closing the Mints to the compulsory acceptance of gold need not affect the existing obligation of convertibility;—the liability to encash notes in gold might still remain. Theoretically this might be regarded as a blemish on the perfection of the scheme. But, for the present at least, it is unlikely that such a provision would compel the United States to deflate,—which possibility is the only theoretical objection to it (p. 200).

It is not only unlikely but unthinkable that deflation would have to follow. The public would not convert their paper dollars for gold, unless they had some profitable use for it, and unless the price of gold promised not only to endure, but to gain. But so soon as the quantity of gold in the market grows, the price of the article will suffer a decline, which will immediately check an incipient tendency to convert. Convertibility is quite another thing when the price of gold is not fixed. Moreover, the necessity to deflate cannot present itself when it is agreed, according to Mr. Keynes's own scheme, that the value of the reserve shall not "bear any logical or calculable relation to the volume of paper money." The deflation here understood is a gold standard phenomenon, which has no place in an Index-number standard. The terms "conversion" and "redemption" have become meaningless and should no longer be employed. To exchange paper money for gold is to buy gold, just as to exchange dollars for a cow is to buy a cow. If one section of the public goes in for gold, another section will thereby be induced to go in for paper money, so that the dollars that are "converted" and returned to the bank of issue will immediately be reconverted and called out again. This process will not deplete the reserve, for the simple reason that the desire for gold is very quickly quenched.

A rather puzzling point is brought up in this passage (pp. 190–91):

> If the Bank-rate and the gold-rate in conjunction were leading to an excessive influx or an excessive efflux of gold, the Bank of England would have to decide whether the flow was due to an internal or to an external movement away from stability. To fix our ideas, let us suppose that gold is flowing outwards. If this seemed to be due to a tendency of sterling to depreciate in terms of commodities, the correct remedy would be to raise the bank-rate. If, on the other hand, it was due to a tendency of gold to appreciate in terms of commodities, the correct remedy would be to raise the gold-rate (i.e. the buying price for gold).

Is it possible that gold should flow out of the country both when (or because) it depreciates, and when (or because) it appreciates in terms of commodities ? Can diametrically opposed causes produce one and the same effect ? The inducement for the business world to export can only arise from an advantage to be gained by it. Gold is exported when it buys more abroad than at home. It runs away when it is threatened by a danger, i.e. when it depreciates on the home market. Let it run, I say. It depreciates because there is too much of it, and there is no other valid remedy than to reduce its quantity. Contrariwise, nobody, surely, would think of giving away the article which is appreciating, of running away from a place and position of safety and vantage. The State need take no preventive measures to stop the outflow of gold when its price, "in terms of commodities," advances. (See below, p. 63, the instance of gold circulating in Switzerland in 1922.) Mr. Keynes's remedy is quite superfluous ; the gold-rate need not be raised, since the buying price for gold has already adapted itself when it is found that gold has appreciated. The depreciation of commodities—which is simply another aspect of the appreciation of gold—is sufficient protection for gold. Gold, in these circumstances, can circulate freely, and will do so, because it must act as money, if it is not to suffer depreciation as a commodity. That is why it does not go to the Bank's reserve, where it would be in the rôle of a ware. By raising the buying price for gold the Bank increases its issue of paper money and so counteracts the falling tendency of the general level of prices ; but gold will produce exactly the same effect if allowed to circulate as currency. In other words : the bank need not interfere at all, but may leave the price of gold to take care of itself. Gold flows out of the country when that is a wholesome reaction ; it stays in—and will flow in—when conditions call for its preservation or augmentation.

How about the arresting of the efflux of depreciating gold by means of a higher bank-rate ? Gold depreciates when prices go up, i.e. when general commodities appreciate.

Of course Mr. Keynes believes—though this belief contradicts an important statement of his made on p. 21 of his book (see below, p. 118 sqq.)—that raising the rate of discount reduces the volume of currency in circulation and restricts credit, the effect produced being a check on prices. With prices of commodities falling, gold would appreciate again and be prevented from flowing out of the country. Such, at least, I take to have been Mr. Keynes's way of reasoning. However, this is to consider gold as money pure and simple, and Mr. Keynes does not consider it so; his gold appears as a commodity; it is not coined, and is not supposed to "buy and sell." Hence gold, rather than regain its price, would seem to be involved in further depreciation. But the remedy proposed simply expresses the fact that gold must be considered as money. How else could gold depreciate and appreciate in terms of commodities? Only money can find itself in this relation to commodities. However thoroughly the scheme of Mr. Keynes purports to demonetize gold, gold reappears as money in almost all his reasonings.

So far, so good. Still I am not satisfied. Raising the bank-rate, more often than it brings prices down, rather forces them up. If, therefore, the remedy here advocated should produce the latter effect, gold would be even more depreciated. The measure proposed is highly questionable. Gold will laugh it to scorn and take care of itself. It will quietly go under cover; cease to be money, and enjoy appreciation as a ware. For gold to depreciate in terms of commodities (as money) is to appreciate as a commodity (in terms of money). Before bank-rate policy can succeed in protecting gold, and the currency as a whole, bank-rate must be more correctly understood than it is at present. And when it is correctly understood it will have become a thing of the past, that is to say, bank-rate will have become a fixed and immutable quantity.

APPENDIX

Why the Price of Gold was Made to Fluctuate by the Unsteadiness of the Economic Conditions

The price of gold is a very modern notion. There was no way of ascertaining it, or rather stating it, so long as no data were available to make up the other side of the equation. These data have been furnished by the computation of Index-numbers. The fluctuations of these Index-numbers are the exact reflection of the fluctuations of the price of gold. If the Index-number went up 10 per cent., the price of gold went down 10 per cent. Each was expressed in terms of the other. So what consideration entitles us, or forces us, to say that the fluctuations of the price of gold superinduced the fluctuations of the price level, rather than the other way about ? It can only be a consideration of cause and effect. To assume that the cause is on the side of gold is to adopt the materialistic view of life, in which gold would play the part of the absolute ruler. It is Silvio Gesell's view, utterly unqualified—and this view will also be found strongly expressed in some of my own earlier publications. I have relinquished it and come to look upon gold as the satellite of man's ways and gyrations.

Very mysteriously the world's supply of gold has from time to time shifted its grazing-grounds. I quote a chance remark of Mr. Keynes's (p. 162) :

> Prior to the discovery of the New World the precious metals were, over a long period, becoming progressively scarcer in Europe through natural wastage in the absence of adequate new supplies, and the drain to the East.

Why this " drain to the East " ? Why did the European nations part with the gold which they so badly needed to carry on business ? It cannot have been the fault of gold, but rather it must have been a weakness of those who let go. The gold having been allowed to escape, it began to be appreciated again ; it appreciated and the price level fell. But something else must have given way previously.

Gold escapes when goods begin to fail and prices to rise : when people squander their substance. It sticks to its ground so long as goods are plentiful. In 1922 gold circulated freely in Switzerland while the export of gold was still interdicted. No one tried to send it abroad, because goods were more plentiful and in less demand in Switzerland than elsewhere. It was not gold that had created this abundance ; gold does not sow the seeds of plenty, but it follows the growth of abundance, and, after having filled people's hands, it escapes from them upon the slightest intimation that the period of growth is passing away. There is no assignable reason why people should not retain their grip on the precious metal ; yet they do let it slip out. It is because they have been demoralized by indulgence ; they allowed the rate of interest to rise, and with it the level of prices, debtors and creditors cheating each other, and ultimately themselves. With characters and hands thus weakened, with populations given to high living and low thinking, the output of real wealth began to slacken, and gold took the hint to be gone—it went because people were too indolent to mind it. Both the boom and the slump were ushered in by deeper lying forces than the chance of gold finds, or the drain of gold. It is to honour gold unduly, or to disparage it unjustly, to attribute to its coming and going the responsibility for the economic tides. I am prepared to trace these unexplained movements to imperfections of the function of interest, but ultimately to the not purely human law of periodicity. That is to say, I have become sceptical as to the possibility of permanent and perfect stabilization ; but it does not imply that I have lost one particle of my ardour for a new and more rational monetary system.

It is no use at all trying to show how the fluctuations of the price of gold might have been avoided. The fact that they were not avoided is sufficient proof that it could not be done. The need of stabilization was not felt by the nations, and what people do not feel the need of, they will not encompass. But it may not be out of place to stop to think whether this stabilization we strive after is really the thing that life demands. Gesell is fully convinced that we might master life by means of a rigidly, ruthlessly stable currency and the tender mercies of an automatic State control. He does not believe that a civilization, or a race, may spend and extinguish itself ; but he thinks that a faulty monetary system can demolish the strongest empires. To me it seems that life is bigger and not quite so simple as it would appear in the materialistic view. Life is subject to the law of change, i.e. of decay. Nations are units, individuals, parts, and therefore doomed to perish. We may attribute their downfall to some such purely extraneous agent as money. But then we ought not to expect to succeed in creating a form of money that will overcome the natural forces of life. A nation that is doomed adopts that kind of money that will carry it to its doom, while a people that is on the rising tide will create a buoyant currency. I have no doubt that a stable currency will lead a nation to a peak of excellence, but I also know that the ups must be succeeded by the downs, and that one nation must decline that another may aspire and rise. I feel it worth while helping to give our Western World another lease of life and so to arrest as long as may be that *Untergang des Abendlandes* prophesied by some. Whether or no the doom is upon us depends on the chances of the money reform. If the nations can muster sufficient initiative and interest to will it, they will by that very act prove that the spark of life is strong in them ; if they fail in this test it will be a proof that they are rotten at the core and rejected by their Mother Earth.

III

MONEY AND WARES

GOLD is a material and a ware; gold, at the same time, is money: the thing that buys and the thing that is bought is one.[1] It may prove a very appreciable advantage to have in our system a thing which will act as a natural and automatic balance weight of the currency. In their measures to control the price movement all the stabilization schemes agree in managing only the money scale in the balance, while leaving the side of wares alone. Of course the method is logical enough, but the question is whether it will work.

Gesell in particular is strict in dividing money from wares. With him things have to be the one to the utter exclusion of the other. He demonstrates the unfitness of gold as the raw material of money from the fact that the gold of the money will turn into a ware to be consumed, used up, instead of being only used, employed, and passed on again. It is surely a very serious drawback when part of the money can be put to illegitimate uses without being properly replaced. But what is it that has been perverting our gold money? Will the inducement to pervert it continue when the standard is managed differently? We have seen what will happen when gold attempts to go on strike: its place is immediately taken by blackleg paper

[1] I wish it to be understood that this interpretation of the case is purely hypothetical. We may at least as reasonably invert the proposition and say that money is that which is not a ware, but the insubstantial reflection of wares. From what follows it will appear, I hope, that gold is never money and a ware simultaneously, but always ceases to be the one when it assumes the function of the other.

bills, which the money employers are able to procure in all circumstances.

The money reform theory is built up on the assumption that the currency is governed by the flow of money, not by the flow of goods; it will not admit that it might be practical also to make use of production in controlling the forces of price formation. Now money itself should really be considered as a ware. Gesell does so consider it, saying in his emphatic way that it is the only thing in existence which is a pure ware (a ware defined as a thing which cannot be consumed by the owner). Also Professor Fisher looks upon money as a ware. "If money buys goods, goods buy money," he says. What is bought is a ware. Why, then, this rigidity in rejecting the idea of a ware which may assume the function of money? Properly understood money and wares are identical, constituent parts of the system of exchange and forming one whole, as the day and night making up the twenty-four hours of the day. If we have not yet come to notice it, it is merely because we have no name to embrace both of them.

I suggest that we ought to profit by this circumstance. If we retain gold as a legitimate money material, we shall have what is money and ware—a material ware, that is, not merely functional—alternately, capable of exchanging at any moment the rôle of one for that of the other, and so of re-establishing equilibrium as soon as it is ever so little affected. In about the same way as the hours between the early and the late daybreak, and those between the early and the late nightfall, play the part of day in summer, of night in winter, so, too, gold—or, more correctly, part of the gold—will now figure on the side of wares, now on the side of money. The equilibrium of the scales will be preserved from within, as it were, by a trick far subtler than anything statistics could perform. Here we have real compensation: variable parts within an immutable whole. Wares are opposed and hostile to wares, as money is to money; we are going to employ this spirit of opposition and enmity in controlling the relation between wares and money. The goods-character of gold shall curb goods and uphold money; its money-character shall curb money and uphold goods. In resisting a plethora of money, gold the ware is the ally and rescuer of money—which would be damaged by plethora; in attacking the plethora of wares, gold the money is the preserver of wares—which, too, would be damaged by plethora. If the goods-scale increases its volume, the ware gold—which has an aristocratic turn of mind—allows itself to be crowded out. Some of it quietly moves over to the money-scale. By this the weight of the goods-scale is diminished, while the weight of the money-scale augments: the result is that the equilibrium remains stable and the level of prices is happily prevented from flinching. The gold which was previously offered as a ware, therefore as a depresser of price—a menace to the dignity of wares—is now offered as money and as an upraiser of price—thus championing the character of wares. The effect is all the more immediate, and all the less is needed to produce the effect. Now reverse the case. If the money-scale tends to exceed, when there is an incipient superabundance of money, part of the gold—gold, owing to its harmonious constitution, is the sworn enemy of excess—allows itself to be crowded out by the paper money. It moves across to the scale of goods; it so escapes from being depreciated, and it saves money generally from depreciation. In this manner the demand for goods is weakened, while the supply is strengthened by the additional gold ware. Gold which a moment ago was demand has assumed the rôle of supply. And so the rise of prices is averted.

An enhancement of the value of money, and so of the gold serving as money, is brought about by the supply of goods growing in excess of the supply of money (or demand for goods): either owing to over-production or under-consumption. All the gold now tends to be money, in order to participate in the enhancement of the value of money and to escape from the loss suffered by goods through being cheapened. Gold will always desert to the camp which promises to win the day. But in doing so it undoes the chances of that camp, because in economic life, as elsewhere, every increase in numbers is made up for by a diminution of value and virtue: more goods reduce the value of goods, a surplus of money impairs the value of money. In deserting to the camp of money gold adds to the supply of money and renders the supply of money more urgent, while it relieves the urgency of the supply of goods; the result being that money fails to gain in value, and the level of prices is preserved from depression.

Money is cheapened, and along with it the gold serving as money, when the supply of goods falls to the rear of demand, either in consequence of failing crops and unsuccessful production, or of excessive consumption. Gold now does not care to be money any longer. It seeks to avert the loss of value; it strives to be a ware, so as to share the gain going to the side of goods. Gold ceases to buy. Consumption is thereby checked, the stocks of goods are replenished, production is stimulated by prospects of really good money, imports are augmented thanks to the improvement of the rate of exchange. The cheapening of money—the enhancement of goods as expressed by the rise of prices—fails to take effect; it does not proceed beyond the incipient stage.

From whatever angle we tackle the problem, our examination invariably leads us to recognize the fact that gold can neither appreciate nor depreciate, in conformity with its true nature, which knows of no differences of quality. In itself it harbours the forces which arrest every movement by curbing it into a circle, which is the symbol of duration, of permanence, and of a stable standard of currency. We are brought to see that, like commoner wares, gold is subject to the menace of depreciation and decay. It is the fact of its being apt for very few uses only that exposes it to this danger. The mere possibility of depriving it of one of its uses affects its status and weakens its position ; for whatever lessens the number of its uses and the scope of its efficiency diminishes, like material decay, its value, and seems to detract from its substance. Decay, this law of laws among things created, is not only material, it is also spiritual. The decay of value, the slow decrease of appreciation, were it only potential, acts in precisely the same way as material dissolution. For gold to be found at fault as a money material, and also found to be easily replaceable, is sufficient to sap the foundations of its traditional superiority. So long as it was considered as the only possible basis of a currency, its power was unassailable. With that conception gone the time of its overbearingness is ended. The threat of replacement hits it like decay. It is shorn of its rays, it ceases to dazzle us. Its value shrinks and it must needs assume a shrinking attitude in the place of its wonted domineering ways. It is forced to shrink and move on, to keep up its speed, to circulate as money ought to. So soon as it tries to slacken its pace and draw breath it is pushed out of the track by paper money, it rolls into the ditch, it ceases to be money, it is degraded, depreciated. All its pride drops away from gold. It can have only one wish : to serve and to prove its worth by serving faithfully and honestly.

I have not space enough here to set forth the principle of depreciation as motive force, which is alluded to in these last lines. Still, in order to make my point a little clearer, I must indicate another form of this shrinkage which forces gold to keep active : interest. He who borrows money, say gold, has to pay interest on it, which obliges him to employ it properly, because otherwise interest will slowly eat up the substance : like decay. He who refuses to lend out his gold, leaving it unemployed, forgoes the interest which he might obtain, and which he can in no conceivable way recover. Under the gold standard, gold hoarders were able to make enormous profits ; when withdrawn from circulation gold might appreciate, not 5 per cent., but 20 per cent. and more. Thus, then, interest acts as a goad both on borrowers and lenders, on those who own the money and on those who need to use it, because not to employ it involves the loss of interest either to him who has to pay it, or to him who wishes to get it. And this loss is a kind of shrinkage, of decay.

We have in gold the most perfect regulator of the currency imaginable. In order to construct a good money machine we depend on three factors, not two only : pure, exclusive money (paper), exclusive ware, and that which is both in one, gold. When assigned its proper place, gold will prove to special advantage that one of its properties which in the old system was most at fault, because it was overtaxed : its sensitiveness—its cowardice, as Gesell denounces it. The more sensitive the automatic regulator is, the more efficient it is. Gold, turned loose into the currency, will do more than the most vigilant statistical bureaus could achieve, in discovering where and when money is short, or money is in excess. A remarkable point about it all is that any quantity, however small, suffices for gold to play the part desired of it. All the gold of all the world is always ready to come to the rescue of the currency of the last and smallest country that adopts the interest standard of currency.

So it will not be necessary for a country that introduces the new system to withdraw from circulation any supposed excess of gold ; but neither will it be necessary to add gold to the circulation or to the reserve in the depleted countries. Gold will see to it that everything is perfectly adjusted. Every citizen is allowed to decide what he is going to do with the gold in his possession ; he may hoard it, he may exchange it for paper money, he can ship it abroad. Nobody will interfere. He can buy as much of the reserve gold as he is able to pay for. The "value of the reserve" is a matter of perfect indifference to the currency. Gold, and a matter of indifference? No one caring? Is it because gold has become useless? No, indeed; on the contrary, gold has only now found its most precious use. But there is this difference that gold is at last free to flow. It will flow where it is needed, and be at hand where wanted. So men's minds are relieved of all care about it. Gold will cease to haunt and waylay and seduce them. Gold will recover its innocence, being removed beyond the distinction of good and evil.

IV

TO WHOM DOES THE GOLD BELONG ?

GESELL does not so much as ask the question. According to him the State can freely dispose of the reserve gold. He even seems to imagine that the public will give up their gold coins for the new paper money, and the gold so obtained he proposes to give away along with the remainder. A fascinating idea: the State receives the gold as money and gives out money, though of another kind, in exchange; but in its hands the gold turns into a ware which that conjurer State can squander away without impoverishing itself. That amounts to saying that the value of the gold is equal to nothing—just nought. And sure enough, if I surrender a number of gold pieces and am given an equal sum of paper money, in exchange for which I can purchase at stable prices whatever I need or desire, I have nothing to complain of: I have received my gold money's worth. I suppose nobody will prevent me from later recovering my gold if I have a mind to do so. However, I shall not do so, unless there is some inducement. My gold, meanwhile, is wrought into some piece of jewellery and presented to a bride on her wedding day. Suppose the bride to be my own daughter: now we shall begin to see somewhat more plainly what these official bounties signify. Had I kept my gold I should have received no money in return. I might have had the gold, at my own expense, made into a trinket for my daughter to present her with. But in that case I should obviously have been at a disadvantage and poorer by the amount of money that I neglected to obtain from the State. In fact, it would have been I that bought and paid for the present, whereas now it is the State that does so in my place. So these gold wedding presents have a value. They who receive them are enriched—at somebody else's expense, no doubt. The scheme of Silvio Gesell is too light-hearted. It plunges us into a pretty muddle, and where there is a muddle there is unfair dealing and fraud. *Quidquid id est,* I fear the State, *et dona ferentem* —especially when it bears gifts.

Mr. Keynes proceeds more cautiously. He realizes that gold is a *noli me tangere*; the vaguer the provisions the better: leave it alone. Professor Soddy has a good many things to say about the ownership of gold which it might be worth while to put to the critical test. He judges more rightly than Gesell in admitting that gold has value, intrinsic, not merely functional value. But he seems to me to be altogether mistaken when he says that paper money "requires no such expenditure" of human effort to provide as gold and silver do. If it costs labour to produce the precious metals out of the bowels of the earth, whereas the printing of token money involves hardly any expense, the difference is made up for by the fact that gold and silver retain all their material value when coined; gold and silver need not be "replaced by the fresh expenditure of a similar amount of human effort" (p. 19), since they have not been consumed. If paper money really cost less to produce than gold money, it could not possibly be as powerful as gold money. But all this will not be properly understood so long as it is not realized that it is not the cost of producing the tangible money tokens that matters, and that confers its special character on money. Money, in all its known or imaginable forms, is merely a reflection of the other side of the balance, the goods scale.

This misconception of the respective characters of gold money and paper money is at the root of Professor Soddy's proposals concerning the management and use of gold under the new system. He writes:

> Since, however, the flight of money means a corresponding contraction of home purchasing power, the Government would do better to print the money that has disappeared itself, and use it to buy the necessary gold for export as opportunity offered, destroying again the money the exported gold purchased.

This seems to imply that the State can buy gold for paper money specially printed for the purpose; in other words, obtain for the thing that costs almost no expenditure of effort to procure the thing that costs a great deal to produce. But I may be misinterpreting the meaning of the passage quoted; at any rate it seems to me to contradict this other statement concerning the purchase of gold by the State (p. 49).

> All immediate new capital expenditure required to initiate the new system, to compensate legitimate interests in banking, acquire the national stock of gold, etc., to be paid for in the new issue of National Debt, yielding interest till redeemed.

Here "capital expenditure," not merely paper money, is required to get possession of gold.

Incidentally the last quotation informs us that Professor Soddy does not consider the State as the owner of the "national stock of gold"—by which he can only mean the gold of the reserve, since it is inconceivable that the State should acquire the gold in the possession of private persons. Now this assumption again raises the question: to whom does the national stock of gold belong, if not to the State? It must needs be to private persons, in the last resort. But who are they? They must be the creditors and shareholders of the Bank of Issue. And by what means could the gold be purchased of them? They do not actually hold the metal; probably they do not so much as realize that they are joint owners of the gold reserve at the bank, and would be greatly surprised if they were approached with proposals to buy their share of the gold treasure. Further there is the question: what kind of value would be handed to them in exchange for what they are asked to yield up? Paper money, money-paper newly printed? I am afraid that it will not do; neither does Professor Soddy suggest that, but says explicitly "capital expenditure" and "new issue of National Debt." Therefore some sort of capital —of course paper capital, i.e. claims on capital and capital revenue; National Debt yielding interest—will have to be offered to the creditors of the bank against—not the gold, which they have never laid eyes on—but the claims on capital which they hold at present, and which confer on them their capacity as creditors. So then it would amount to an exchange of claims on capital for claims on capital, the ones exactly as good as the others, since it is to be a fair deal. And the gold at the Bank would not have changed owners, *because there are no owners*. The gold at the Bank is not a corpse, as I put it above, but a ghost, immaterial, unreal and unrealizable. This I am going to demonstrate by an examination of Mr. Goldsborough's Bill. It is provided that—

> The Secretary of the Treasury shall divide all the gold . . . into two parts, one part to be known as the reserve against outstanding gold bullion dollar certificates and equal to 50 per centum of the value of the gold certificates then outstanding, and the remaining part to be known as the surplus in excess of the said reserve. This remainder or surplus shall be forthwith transferred to the general fund of the Treasury as the initial profits of the new system.

This clearly signifies two things: namely, that Mr. Goldsborough considers the gold of the reserve (*a*) as the property of the State, and (*b*) as an asset. It will be necessary to try and visualize the process by which the State might be imagined to realize the "initial profits of the new system." For surely a profit it is only if it can be realized, made something profitable of. The new system provides for a sufficiency of money circulation, and guards against an excess of it. Now suppose the Treasury to begin to meet its liabilities by selling the demonetized gold and employing the proceeds for its payments. A crowd of strange problems leap up at the suggestion, of which I will indicate only a few.

(1) Who will profit by this novel method? For it does hold out the promise of some advantage. Surely the taxpayer, since what the Treasury realizes out of the sale of gold need not be raised in the form of taxes. But out of whom does the taxpayer make his profit? There is only one taxpayer: the nation as a whole; the owner of the gold is this same nation; so the taxpayer can make his profit only out of himself.

(2) Who will buy the gold? Gold is demonetized in the United States; it is not supposed to be used as money. Of gold for industrial and artistic uses there is enough in the market. Consequently the likelihood that the American public will absorb the Treasury's gold is very small. However, let us suppose that it will acquire the gold. These supposed buyers of gold acquire the yellow metal as a ware, either for use or for a speculation. This accretion of gold in the market is tantamount to an increase of gold output; for the gold appears as a ware, not as money. To keep the level of prices stable, an equal increase of the money circulation must be effected. By all the stabilizing schemes this necessitates a new issue of bank-notes, and by Mr. Goldsborough's scheme this would necessitate the augmentation of the reserve. So the gold only just disposed of would have to be repurchased.

However, that is not all. We assumed that the Treasury will spend the money received for the gold to meet its liabilities. The volume of paper money, therefore, is not diminished. If a few weeks later the Treasury can order new notes to be printed to repurchase the gold, it will be as if the Treasury had paid a debt with mere money-paper, in the same way that was practised by the Governments of civilized nations (within all our memories), and with the same result: the impoverishment of a section of the community through inflation. I need not say how such methods should be judged. A disturbance of the price level is to be created ere yet the stabilizing measures have had time to take effect!

"The surplus shall be forthwith transferred to the general fund of the Treasury," such are the words of the provision. It would be amusing, if it were not pathetic, to find the author of a Bill to stabilize the purchasing power of money imagining that large sales of gold would have no effect on the level of prices, i.e. the purchasing power of money. And have we not seen Mr. Keynes also venting the same view? We must pick away at this tangle until we have unravelled it. It is not from a cavilling spirit that I am trying to hunt down errors. In this critical part of my inquiry I am bound to clear the ground as best I can from whatever misconceptions I find to be lurking in the proposals of the reformers.

Mr. Keynes suggests that to get rid of the gold excess the Mints should offer gold cheaply. Mr. Goldsborough wants to sell half of the American reserve, which is an enormous quantity. If it could be done—but it cannot—this is what would follow. Extensive purchases of gold will only be induced if the article is supplied cheap. But this will reduce the value of the gold retained and "known as the reserve," which ceases to be "equal to 50 per cent. of the value of the gold certificates then outstanding." Strangely enough a change of the value of these certificates must ensue to complicate matters further. (We ought to begin to realize, by this time, what an elusive thing this value is!) The paper money would appreciate. In buying the gold of the reserve the buyers pay with paper dollars, which is the only legal tender. Even supposing that these dollars were passed on again, the first effect would be to reduce the circulation and to depress the level of prices:

the money which buys gold, as a ware and in large quantities, is not available for the purchase of other goods, which lose in price in consequence of failing demand. Thus the purchasing power of the paper money, in terms of ordinary goods, grows ; stability goes overboard. The value of the gold reserve shrinks all the more, and the only remedy will be to buy up again the gold that has only just been sold. The yellow metal comes home to roost.

Any kind of interference with gold is likely to depress prices at first, because it shatters credit, which means a destruction of purchasing power. Credit will recover more or less speedily. By our present hypothesis the volume of legal tender is not diminished, since it is assumed that the Treasury does not cancel the notes received in payment for its gold, but passes them on to its creditors. This is to favour these creditors : they are given money at a time when it is appreciating, and they will be all the less likely to pass it on in a hurry. Others are at a loss for it, and by their embarrassments compelled to dispose of their dispensable property, which can only be their cheating gold. They sell the gold, and that amounts to saying that they buy with gold, demonetized though it be. Gold thus acting as money, there must result an excess of demand, with a rising tendency of prices : the price movement, from a falling one, swings round to a rising one. Stability is impossible so long as Treasuries can sell gold as if it were their property.

I hope that we have been brought to understand that the gold of the reserve is not an asset of the State to be disposed of at pleasure. Public debts cannot be paid with the national stock of gold. The sale of this gold, no less than the giving away of it, involves unfair dealings with certain sections of the population, some being benefited, some injured. The expected gain and " initial profits " of the Treasury resolves itself into nothing plus dishonour.

The error of considering the precious metal backing as

a real value and an asset of the Bank of Issue, is also to be found in Mill's *Principles of Political Economy*, Book III, chapter xii, § 2:

> We just now made the imaginary supposition that all persons dealt with a bank, and all with the same bank, payments being universally made by cheques. In this ideal case, there would be no money anywhere except in the hands of the banker; who might then safely part with all of it, by selling it as bullion, or lending it, to be sent out of the country in exchange for goods or foreign securities. . . . There would be in all this nothing to complain of, so long as the money, in disappearing, left an equivalent value in other things, applicable when required to the reimbursement of those to whom the money originally belonged.

It is the same conception as in the schemes of Gesell and Mr. Goldsborough: when gold that was money is no longer directly employed as money, it can be sold at a price. But gold is not like an iron machine which, when scrapped, can be sold as iron; gold is not a machine at one time and sheer metal at another time, but always the same: just gold. To sell gold is to depreciate gold, and not only what quantities may be offered for sale, but all the gold in the possession of people along with it. What the sellers realize is abstracted from the property of the owners of gold; it is mere loot. The case which Mill supposes is exactly that of the Bank of Issue, the bank-notes being the cheques in question. But the gold of the reserve is not an asset of the bank; it is an asset of those who hold the bank-notes which are issued against it, and "to whom the money originally belonged." The value of this gold has been emptied out, as it were, into the bank-notes, and what the reserve contains is an empty husk, worthless. To sell the gold at a price is to rob the holders of the notes; it is to dispose surreptitiously of what has been given in trust. When I find that the chief errors that I can detect in the writings of the money reformers are the same as those that mar the *Principles*, I realize what little progress has been made in the understanding of money and the problems raised by money. It is no wonder that the world is unable to extricate itself from the tangle in which it has become involved. How could the problems of debt, both national and international, contracted and imposed, be solved when the highest experts do not know what are the assets and what the liabilities !

(3) There is an even graver aspect to the problem. Gesell rightly points out that selling the gold reserve to the highest bidder would depress the price of gold and embarrass other nations. For the sake of the argument let us assume that the United States Treasury pay its foreign debts with shipments of the golden profits of the new system ; it is clear that owners of gold, no matter in what shape, the world over are hit below the belt by these gold exports, which, by increasing the supply of gold, cheapen the price of gold. Far-reaching political consequences would disturb the peace of the world. Nor would the disturbance fail to affect the originators. Not only the gold outside the United States is cheapened, but the American gold likewise. That part of the reserve which is being maintained loses in value and becomes insufficient as compared to the paper money issued. This depreciation may easily take as much out of the gold in American hands, either public or private, as the exports abroad amount to, so that the payment of the debts in question would in reality have been made out of indirect taxation of the American owners of gold.

We need not carry this argument any farther, for it has become clear that the "initial profits of the new system" are nowhere to be discovered. They are purely imaginary, the Fata Morgana of a hazy notion of the gold problem. The gold of the reserve has no assignable or calculable value. If you halve it the value of each half, used separately, is not to be compared to the value of the whole. These halves will be as worthless as half a watch or half a gun, or half a pair of trousers. Gold, all the gold of all the earth, should be considered as an indivisible whole, and absolutely unmanageable. Schemes involving any form of State control are doomed to failure. In their subjecting gold to interference by the State resides the vice of the proposals which we are reviewing. Their views concerning the ownership of the gold of the Banks of Issue may be summed up thus: Gesell wants to *throw* the whole of the currency gold *away*, and flatters himself with thus having controlled it out of the currency.[1]

Mr. Goldsborough wants to *sell* half of the currency gold, while keeping the remainder.

Professor Soddy proposes to have the State *buy* the stock of gold and entrust the League of Nations with the control of the national stocks.

Mr. Keynes advises to *keep* the national stock of gold buried (at least in Great Britain; the United States are advised to sell the excess of their gold).

Mr. Lowenfeld forgets its existence, and so gets rid of it by ignoring it.

This conflict of conceptions is a sufficient reason for nations and parliaments to fight shy of the money reform schemes. Some degree of unity must be reached before the idea can be put in practice. The result of the present inquiry is to the effect that:

(1) The gold of the national reserves is not an asset and cannot be given away, or sold, or acquired by the State.

(2) The gold must not be controlled, but allowed to move freely through the markets, both nationally and internationally.

(3) The price of gold must always coincide with the monetary unit of a currency once stabilized, so that the

[1] An English writer on currency reform, Mr. Arthur Kitson (author of *Fraudulent Standard*), in a recent article deriding the efforts to raise the gold gone to the bottom of the sea in foundered ships, reiterates his earlier proposal to make an end of the gold nuisance by sinking all the currency gold in the most inaccessible part of the deepest ocean.

gold dollar (gold sterling, gold mark, gold franc) cannot be a variable quantity of gold.

(4) For practical purposes it is desirable that a portion of the gold should be coined and allowed to circulate; in other words, gold should be recognized as legal tender to an unlimited amount.

V

SOME AFTERTHOUGHTS

A. Why There Must be Gold Reserves

After all that has preceded we may at last find a plausible reason for the existence of and the general belief in gold reserves. Neither the defenders of the faith in gold and the gold standard, nor the scorners of this faith, are agreed among themselves in their interpretation of gold reserves. Professor Fisher and Mr. Keynes hold views which are widely different, nor does Professor Soddy, though he is as unacademic and as radical as Silvio Gesell, take up the same attitude as the German reformer. The commonest opinion is, or used to be, that the gold reserve is necessary as a backing for the notes issued. Notes, it was maintained, must be convertible or redeemable, without which security they would not circulate. This notion has been thoroughly exploded by recent practical experiments, which have proved that notes circulate the more readily the less they are weighed down by gold. Another idea has in consequence of the disruption of the old theory found fairly general acceptance: that a gold reserve is needed as a security against untoward emergencies. I have discussed this conception in my criticism of Mr. Keynes's "war-chest" plan. In normal times the reserve is not needed, while in an emergency gold is the last thing that a government will part with. So this explanation is not likely to go a long way.

Gold accumulates naturally and inevitably in a country which manages to keep its house in order. A very notable example of how it is brought about is furnished by the vicissitudes of the Argentine currency. A gold standard in connection with a bank of issue was established in 1899. There was no gold to begin with. By April 1900 the reserve amounted to 26 million francs, but by July it was empty again. It continued mostly empty all through 1901 and 1902, which did not, however, prevent paper pesos from circulating merrily, nor business from doing excellently. By February 1903 the reserve was at 20 million francs, by the end of the same year at little short of 200 million. From now on till 1914 the reserve kept on growing with hardly any setbacks. In 1912 it went above 1,000 million francs.

Now this heap of gold must be considered as accumulated wealth. It represented the surplus of exports over imports —production over consumption—within a period of some twelve years. Gold was handed in by merchants as soon as and so long as the paper peso was better money than gold, so that the foreign buyers of Argentine exports preferred to pay in gold. I get the figures here quoted from a disciple of Silvio Gesell's, that is, a sworn enemy of gold reserves and gold notions generally.[1] His enmity towards the institution blinded him to the fact that in no other way could Argentina have acquired and saved this wealth. If the Argentine people had not been able to accept gold in payment they would have been forced to export less, or to import more goods for immediate consumption. Wares for consumption are very fine and desirable, but they cannot be saved; they will perish. The Argentine producers would have been obliged to consume what the acceptance of gold enabled them to save and lay by. Of course that is only one aspect of the case. In the long run the nation might discover a serious disadvantage in laying by gold rather than consuming the whole of its production. Indeed, a point of time must finally be reached

[1] Ernst Frankfurth: *Geldbriefe vom Silberstrom*, Montevideo, 1915.

when this economizing cannot be continued, because it would result in nothing but the piling up of a hoard of utterly useless gold. However, that is not our concern here. What I wish to stress is that within twelve years, 1,200 million francs' worth of gold collected in the Argentine reserve. It happened quite naturally, freely, as though nothing else could take place. The money house had been put in order, the country was thereby given a respite from civil strife and a chance to work and develop; all these favourable features would have been inhibited if the people had not been able to save, and to save in the shape of gold. In much the same way I imagine gold reserves to have formed in the origin before there was any doctrine about them at all. We may be quite sure that the practice of gold reserves preceded the theory, nay, established itself in the teeth of some theory opposed to it. Gold reserves were not scientifically devised and deliberately planned; they just happened to come into existence, the way all really good, that is necessary, things do. They are necessary, not for reasons of security or national safety, but simply because they cannot be helped. The nature of men and the properties of gold combine to make gold reserves spring up and grow, as warmth combined with moisture will bring forth plant reserves.

Beside the national reserve there are the private hoards of gold, and not a negligible quantity by any means. During the war the reserves of the Banks of Issue were doubled and trebled. Much of the gold that had been kept hidden in private treasuries was drawn to the national vaults. Six years after the end of the war there is hardly any gold in circulation. It seems impossible to bring this precious gold out again. Gold circulated profusely in Switzerland all through 1922; by the end of 1923 it had disappeared again without any measures having been employed to stop or discourage it: private "war-chests" had been formed again. Gold cannot be ordered about. It will treasurize, and that being its innate characteristic it may be advantageous to allow national reserves to exist, rather than have it collected in private hoards only. Silvio Gesell has a great deal to say about the menace arising to the stability of the standard from private reserves of money. However, this is true only so long as the flow of gold is impeded by gold standard regulations. We shall come to understand yet how this menace may be averted by means of a proper regulation of the rate of discount. Far from being denounced as a danger, private gold hoards are more likely to be appreciated as a safeguard to the currency. They should be accepted as a natural growth, as things of nature. We must try to discover their uses and suit our policies to the conditions which they create and impose.

Gold reserves are stores of power, of energy. We cannot consume gold, but gold has that quality in it which will create consumable goods at a moment's notice. The goods themselves it would be impossible to store up. By storing the gold we reserve the capacity to produce. I should think that this is the way of nature. Nature and life require a store of value—value meaning power, energy—and cannot afford to have the whole of their resources in actual use. One half of every whole wants to be lying at rest. Man's life is divided between activity and rest, giving out and taking in energy. Nature has her seasons, the Earth has her climates and zones, her gardens and her wastes, her arctic ice to make up for her tropic exuberance. The economics of man are a copy of the economics of Nature. Man can employ no more than one half of his means, the useless half—the buried reserves—constituting the necessary condition of the usefulness of the other half. The Chinese Laotse has beautifully expressed this in the simile of the potter's art: "Men take a lump of clay to shape it into a pot; on the nought, the void, of it depends the vessel's usefulness."

The idea of Silvio Gesell, so strongly stressed, that in a

good currency all the money issued must be in perpetual motion and activity, with no reserves either public or private, is a very grave error, and contrary to Nature. It is impossible. What would happen if the attempt were made is this: all the gold in its demonetized state would constitute itself as a reserve, ever ready to become operative, to move on into the line of action. Nay, more; it will be most active when most strongly repressed. As it is impossible to force both halves of a whole into action, it is also impossible to force both halves into inaction. The idea of ruling out gold may be talked about; but the first move towards its application will be broken on the rocks of necessity.

There were reserves of gold, both public and private, before the rudiments of a theory of currency or banking had been evolved. In Pepys's Diary I found this piece of information (November 23, 1662):

> There is also this week dead a poulterer, which was thought rich, but not so rich, that has left £800 per annum, taken in other men's names, and 40,000 Jacobs in gold.
>
> (July 11, 1666): I shall get in near £2,000 into my own hands, which is in the King's, upon tallies; which will be a pleasure to me, and satisfaction to have a good sum in my own hands, whatever evil disturbances should be in the State; though it troubles me to lose so great a profit as the King's interest of 10 per cent. for that money. (See also May 19, 1663.)

When, in 1798, French troops took the town of Berne, they carried off the Bernese Republic's gold reserve, a very considerable pile, which the thrift of prudent rulers had collected. Gold does not change its properties, nor man his propensities. Therefore men will create gold reserves, which we may call a store of value, or by any other name. But it is futile to erect far-fetched doctrines and devise elaborate schemes for the management of gold. Mr. Goldsborough's fifty per centum plan is artificial, and so is Professor Fisher's dollar-for-dollar suggestion. Of all the schemes proposed

I prefer Mr. Keynes's, which leaves the proportion of legal tender and reserve to find itself. It will find itself. All the gold that can be profitably employed will flow out, and whatever gold finds no suitable employment will flow in. Laws to regulate the movements of gold are only a lure and a deception. No nation can dispose of its gold at its own pleasure or according to its own laws, because the action of one State reacts on all the others, so that a situation is sure to arise in which the law has to be either revised or circumvented.

All the gold of all the Earth forms one whole, and of this whole no more and no less than one half, or one side, can be active at a time. The other half constitutes the reserve, and it makes no difference whether it is privately owned or collectively. We may consider even the undug gold as part of the reserve. However deeply hidden, however inaccessible it may seem to-day, the force of circumstance will bring it within the reach of men that need it. For gold is the essence of energy, although by its chemical constitution it may seem the deadest of materials. Its alchemy is determined by the hold which it has on the spiritual forces of men. Merged in a world where all that he can handle is perishable, man clings to the one thing which seems to be exempt from decay: gold. This gold, which in itself displays no energy, moves the energies of man and so constitutes the source and reserve from which flows what man creates for his uses.

In all times war—with revolution in its wake—has been the disperser of gold reserves. An unheard-of situation has been created by the late lamentable peace treaties. One of the most energetic and best-equipped nations has been forced to reduce to a fraction its expenditure for purposes of warfare. The inevitable consequence, even in spite of Reparations (which will crush the recipients), will be the accumulation of gold in Germany. What the other nations spend for armaments the Germans can lay by. Perhaps they will save in the shape of foreign investments; but quite certain it is that a stream of gold will pour into Germany and fill reservoirs, which the fear of other nations will call war-chests, and look upon as purposely planned and made. Nothing of the kind, but a sheer natural necessity. If you destroy a nation's armed defences, some substitute organ will develop, a spiritual defence with a spiritually minded nation, a golden armour with an industrially and commercially minded nation. Whichever the German nation may prove itself to be, its defenceless condition will make for its enrichment, and the great problem for Germans—and not only for them—will be to find an outlet for the wealth that flows from organized industry, for the energies accumulated in a store of gold.

I was considerably interested in Mr. Lowenfeld's appreciation of gold reserves. Like myself he introduces the notion of energy, in the scientific or philosophical acceptation of the term, into the discussion of the problem. I quote a few passages from the chapter entitled "Money in Fetters," p. 65 sqq.:

For if we consider reserves to be accumulated energy, then currency represents it, and this energy can be used for any purpose by means of the most pliable of agencies, this very currency.

The currency itself merely acts as an agent, therefore its outward expression need be no more than serviceable, and as little energy as possible should be expended in its production.

As a consequence we selected gold to represent money, an article of value raised at great expenditure of energy which cannot be liberated again. In fact, we failed to realize that currency must stand for an accumulated store of goods, and need by no means itself be an object of value.

The diamond consumes energy in its creation, and gold absorbs man-power when it is produced. Neither effort can be liberated and used again, and this qualification our accumulations must have if they are to furnish us with life's necessities, of which gold is one of the least important.

Energy has, in fact, to be expended before this guarantee can be met; it is therefore the very opposite of a store of reserve.

Instead of following this precept of evolution we nominally store gold, a material which has absorbed energy in its collection and is incapable of liberating it.

This author, then, is very insistent in contradicting my interpretation of the case as stated above. If Mr. Lowenfeld's view of gold and gold reserves were right, my whole system of currency regulation would collapse. For should it be true that the energy stored up in gold reserves cannot be liberated—having thereby ceased to be energy at all—the use of gold in a currency would be sheer foolishness; also it would be foolish to suppose that the demonetization of gold would make any difference to anybody and call forth any sort of manifestation. Now I suggest that Mr. Lowenfeld persuade the British Government to demonetize gold and then watch the energies liberated by this menace to the value of gold and gold reserves. He would be amazed at the fierceness of these energies. Professor Soddy thinks that the United States would go to war over its gold reserves! Nothing, I think, could more forcibly convince our critic that the energies of gold are ready enough to be called into activity.

Even apart from this special consideration it is not difficult to demonstrate the way in which the economic energies of gold are displayed. Take the case of the formation of the Argentine reserve discussed above. I mentioned that this reserve constituted wealth economized by the Argentine people. It seems so, but curiously enough, on a closer examination of the case, it appears that this wealth was not saved, but spent. Consider it: to purchase gold surely is to spend one's means. If you want to economize you have to refrain from purchasing altogether. It was as if the Argentinians had purchased some other kind of commodity, of a consumable nature. For by exchanging their wheat and meat for gold they employed as many workers as were needed to produce the quantity of gold stored—which must have been exactly the same quantity as was required to produce the wheat and meat which they gave in exchange for the gold; for surely the gold obtained by a given quantity of labour must be equivalent to the price of the commodity in the production of which an equivalent quantity of labour is spent. By purchasing gold they did not economize in the fashion of him who shrinks from spending, and so narrows the range of employment. For the very reason that labour was expended to accumulate that gold, the reserve so formed actually represented a store of labour that might be liberated when called for.

The gold reserve, as it was formed through the sale of consumable goods, might be spent again in the purchase of consumable goods. When this takes place labour is drawn away from the production of gold to be directed and applied to the production of the consumable goods that have come to be required. Is not this the liberation of stored-up energy? Most obediently the labour stored in the gold reserve will turn out to bear at the point where its services are needed; its energy is made to become operative when and where desired.

Something has already been hinted concerning the competition between the dug gold and the undug gold. Where there is competition there is a display of energies. If gold should cease to be demanded in quantities sufficient to keep the output of fresh gold at its wonted level, gold would become so much cheaper that it would not pay to work the mines any longer, and, the mines being closed down, the labourers occupied in them will represent energy liberated by the stored-up energy of the gold reserves. And this is the reason why: the gold reserves are in a position to supply gold cheaper than the mines. If it were not for these reserves, the situation would be different: the demand for gold would be more urgent; it would not be possible to close down mines, and no fresh labour would become available to supply new demands. On the other hand, of course, new demands would be prevented from being formed and from manifesting themselves, because the demand for gold would occupy their place.

I am afraid that Mr. Lowenfeld is somewhat unphilosophical, in spite of the inspiration drawn from Monsieur Bergson. His ideal of a reserve, of a store of energy, is one in the production of which " as little energy as possible " has been expended. That would be cheating nature of her dues. The law of energy has it that by no trick of cunning can you succeed in getting out of a system more energy than has been put into it. The suggestion of cheaply got money reminds me of certain observations which as a traveller in Germany I made in 1923, when money was produced in a way which was supposed to cost little expenditure of labour. If the labour was not expended in the manufacture of the money, it certainly was not saved, but expended in doubled and trebled force in the increased difficulties of ordinary trade.

I am obliged to Mr. Lowenfeld for having given me an opportunity of touching upon these questions. My currency stabilization scheme is, as a matter of fact, derived from a study of the workings of energy in economic life, and the title of my German work is to be: *Die Philosophie des Zinses: Versuch einer Energetik der Volkswirtschaft.*

B. How Much Gold is Needed?

Mr. Keynes more than once expresses an opinion that there is an excess of gold in existence. Professor Gustav Cassel, on the other hand, rather inclines to an apprehension lest the gold supply might prove insufficient, so that the world is likely to suffer from a chronic shortage, because the production of gold can no longer keep pace with the expected expansion of business. To me it seems that with a stable currency the volume of gold available for monetary purposes need not vary at all. If the quantity required has up to now steadily increased, it was not on account of the expansion of business, but the progressive rise of price-levels. Looked at from the other side, this rise was a depreciation of gold. The quantity of gold supplied grew, but its efficacy was diminished, so that by the time when the number of weight units was doubled the power of every unit was halved. The expansion of business does not depend on the volume of gold, or currency generally, expanding at the same ratio. Such expansion creates of itself the necessary money through accelerated circulation. Expansion proceeds from readier consumption; consumption returns the money, absorbed by production, to the current. The gold piece or bank-note changes hands twice a day instead of once, and it is as if the number of money tokens had been doubled. This process may be carried on indefinitely through credit. What has come to be termed "credit"—very inappropriately—is nothing but an aspect of the speed of money. Gold may be said to circulate at an infinite ratio when it lies still, a host of substitutes, more or less visible, performing its functions—vicariously, if you like Professor Fisher's interpretation.

Thus there are reasons for supposing that, under a stable standard of value, the currency will not absorb more new gold than is needed to make up for the inevitable losses and wastage—and to respond to an increase of population. The production of gold may be diminished in quantity; but the usefulness, the efficiency of the gold already produced will no longer keep shrinking and dwindling away through depreciation. The invariable price of gold will be translated into an invariable quantity of gold—for every head of the population—to be employed in the currency.

The production of gold is regulated by the real demand for gold; the demand, in its turn, by the need of gold. And it would be foolish to distinguish between wise and unwise needs. A craze for gold, when it develops, is as natural as appetite for food; it springs from the unexplored and ungovernable deeps of our mass-instincts, and responds to needs which are not for gold. Gold, the useless one, that does not feed nor clothe, nor even delight us, is sought after when men's energies take a certain turn and wax beyond the satisfaction of purely material needs. But the law of compensation takes care that when one section of humanity tends to accumulate gold, other sections are fain to let gold go. The getting of gold, now as ever, requires an extra effort: labour, enterprise, danger, adventure, invention. Men will get gold when their spirit of enterprise and invention and adventure, their zest for danger and labour is not otherwise satisfied. The commercial use of gold is not much more than a side issue. Therefore the monetary systems will not absorb more gold than what is dispensable, but evolve substitutes whenever circumstances call for them—e.g. when little gold is found, when gold is hoarded by the public, when much gold is used industrially. But it is clear that if gold is bound and fettered into the system by static regulations after the manner of the gold standard theory, the necessary adaptations called for by changing circumstances can only be brought about by violent ruptures, such as wars and revolutions. A dynamic and flexible, pliable system which leaves gold unregulated removes these risks by keeping the access to the gold reserves open.

Some of these questions are discussed once more in the last part. The conclusions here arrived at seemed so amazing and unique that I felt it necessary to produce as much evidence as would come to hand. After six months more I have collected such a mass of further proofs that I could easily compose a new chapter. I shall reserve this task to a book in which I intend to present the case in a philosophical, rather than a practical, strain.

PART III

THE UNSOLVED PROBLEM OF INTEREST

I

CURRENCY AND INTEREST

ANYONE that seriously considers the currency problem is bound to encounter the problem of interest. With the exception of the Goldsborough Bill all the writers whom we are considering touch upon it. Silvio Gesell has derived from his insight into the nature of money a theory of interest quite of his own, and along with it a stupendous programme of económic and social reconstruction. Considering, as he does, interest as exploitation, extortion, a wrong, his genius as an observer and thinker has been perverted to the meaner uses of the agitator and revolutionary. He may have influenced Professor Soddy, who also looks upon interest as usury, and debt a device for depriving the workers of the fruit of their labour. He, too, holds interest to be the direct outcome of a faulty monetary system. He writes (p. 34):

> For interest, once the distinction in question is grasped, appears as the simple and natural result of abstinence from publicly coining sufficient money to circulate the revenue, to keep prices and therefore our mutual indebtedness constant. If prices are held constant that is an absolute proof that the community can meet the demands made upon it for wealth, and from such a community interest on money loans could not be asked.

Professor Fisher and Mr. Keynes, though standing firmly on "capitalist" ground, have not missed the social import of the reform. Professor Fisher, after pointing out that the rate of interest will have to be had recourse to in regulating the influence of credit on the movements of price, points out, as the chief advantages gained by the reform, that it is the "path that leads away from Bolshevism, populism, and socialism, because it leads away from the automatic injustices which beget these immensely dangerous quack remedies." In Mr. Keynes's book we are struck by certain powerful passages in which the dangers of "compound interest" are strongly emphasized. Let me quote rather than epitomize (p. 10):

> Those secular changes, therefore, which in the past have depreciated money, assisted the new men and emancipated them from the dead hand; they benefited new wealth at the expense of old, and armed enterprise against accumulation. The tendency of money to depreciate has been in past times a weighty counterpoise against the cumulative results of compound interest and the inheritance of fortunes. It has been a loosening influence against the rigid distribution of old-won wealth and the separation of ownership from activity. By this means each generation can disinherit in part its predecessors' heirs; and the project of founding a perpetual fortune must be disappointed in this way, unless the community with conscious deliberation provides against it in some other way, more equitable and more expedient.

This very searching analysis—and who does not feel in it the thrill of a certain restrained passion?—expresses the social demands of the liberal economist. Mr. Keynes finds the process described inequitable and inexpedient, and we agree with him. But at the same time he finds the result of it not only beneficent and expedient, but also just, and we certainly do not disagree. All the more regrettable it is that Mr. Keynes does not examine, nor seem to be concerned at all, with what will happen when, thanks to a perfect and stable currency, the periodic relief of the burden of debt and compound interest—which relief he represents as a natural necessity—is not brought about automatically and forcibly. Sure enough he advocates the Capital Levy to reduce the inequality. But this measure can only have a meaning in a State which is excessively in debt. But how is a State to contract unmanageable debts when, owing to a stable currency, its economic life flourishes? Mr. Keynes has not considered the problem of interest in a stable currency; there is in his book not a word as to what will become of interest and compound interest.

In order to enable the reader to understand my conception of interest I am compelled to demonstrate Silvio Gesell's, from which mine is derived. Gesell expects of the currency reform a powerful influence on interest. His proposition is to this effect: A stable currency creates the very most favourable and most durable conditions for the general prospering of the economic life of a nation. The unemployed find employment, the growing demand for labour raises wages—the earned income—all along the line. As, owing to the stability of the level of prices, the rising wages cannot force up the cost of living and so be reduced by their own increase, the sums needed to add to the earned incomes must be found elsewhere, outside the current of prices. To-day the total proceeds of labour are divided in about equal shares among the earned incomes and the unearned incomes. If the one side is to grow, it can only do so at the expense of the other. Thus the anticipated gain of the earned incomes must reduce the unearned incomes, and hence a stable currency may be expected to curtail interest. And furthermore: the recipients of unearned income being weakened by their steadily growing losses, while at the same time the workers are strengthened and rendered more confident, the age-old class war will be decided in favour of labour. Capitalism and exploitation will be ended. This goal and final liberation will be achieved, when a stable currency will have been established among all the nations of the earth: in the universal ocean of wealth arising from the overthrow of the barriers of productive work, interest, this incubus bred of penury and necessity, will be overwhelmed and drowned. The same process can be looked at and accounted for from another angle yet. Interest is paid, Gesell points out, because demand for capital is bigger than supply of capital. Now a stable currency, in liberating the forces of production and moreover bringing in its wake innumerable improvements of method and organization, must inevitably alter the traditional ratio of supply and demand. Supply will grow and swell beyond the point where it can be counterbalanced by demand; consequently interest will decline, and the more it declines, the more powerfully will be displayed the forces which reduce it.

We shall come to see that Gesell's arguments proceed from a fallacy. Yet they are greatly to his credit and a testimony to his penetration. They are an expression of the fundamental fact that interest is the governing factor in the circulation of money. It was from the recognition of this fact that Gesell was able to evolve his theory of a dynamic currency entirely liberated from the static conception underlying the gold standard doctrine (and the hybrid conception of Professor Fisher, who still considers the purchasing power of money as vicarious, i.e. borrowed from the gold of the reserve). As, so Gesell argues, with a permanently stable level of prices the rate of interest is bound to decline, a point will before long be reached when the owners of money will cease to lend their surplus. This causes a slowing down of the circulation. The money administration, to preserve the level of prices, will be forced to issue more and more new notes, a process which will be continued until finally all the savings of the people will be in the shape of money hoards, the total amount of outstanding bank-notes exceeding all bounds. What is to be expected of such a state of things? At the slightest scare the public will rush to market to convert their money savings into real goods and commodities, and at one blow the stability of the price-level will be overthrown. Therefore it is futile to try and inaugurate stabilization without forestalling such an eventuality by altering the very essence and character of money itself. Money must be so constituted as to be debarred from lying idle. Money must be made consumptive, dying: *Schwundgeld*, waning money, decaying money, shrinking, shrivelling money, self-consuming money. The traditional gold money failed to stand the test because it does not conform to this requirement; the money of the new system must suffer a regular progressive depreciation, corresponding to the depreciation that goods suffer from their perishable material or functional constitution. The individual bank-notes are depreciated by $\frac{1}{1000}$ of their face value weekly, i.e. 5·2 per cent. per annum. In order to be maintained at their full value they have to be stamped weekly by the holders. The owners of cash thus become liable to a loss, which they can only avert by getting rid of the money through the acquisition of some equivalent that is not cash. The result is that, thanks to this pressure on the holders of money, the circulation becomes absolutely regular and constant.

(Such perfect regularity would in itself guarantee a perfect stability of the purchasing power of money, without any further measures. However Gesell himself refused to admit this consequence when, in one of the earlier stages of my drifting away from his system, I brought it to his notice.)

I must beg the reader's leave here to express my opinion that the idea of consumptive money, as sketched above—alas! too summarily—is one of the most fertile and penetrating discoveries ever made. Practically quite useless, it is the "only plumb-line for the construction of a true theory of money," to use one of Gesell's own felicitous phrases. It is a dazzling and bewitching invention. Gesell himself has fallen a victim to its fascination, so much so that he has become incapable of sounding and scrutinizing it further, lest some fallacy might be at the bottom of it. I will here add the confession that it was this striking and catching idea of shrinking, decaying, and therefore *living*, money that took hold of my imagination and won my allegiance to Gesell's reform. I too was dazzled by it—but also fired by it. The conception of consumption and decay as being the principle of life itself became the dominating object of my thought and study. I found out its universal application, and little by little it was revealed to me that the shrinkage and decay of money is not so much a material as a spiritual fact operating through the contrivers and users of money: men. Thus I learned to dispense with *Schwundgeld* as a practical requisite of a stable currency.[1]

In a previous chapter I threw out a hint as to interest being the true principle, or rather manifestation, of the decay and consumption of money. I must now enlarge upon this theme a little. Consumptive money forces its holder to spend it quickly. Now he who spends money makes an effective demand—for goods, if he goes to the shops and markets ; for real estate and capital, if he acquires

[1] The idea of the shrinkage of money is to be found in the writings of the more progressive French economists. I quote from Jean Labadié, *Si j'étais ministre des finances*, Paris, 1922, pp. 90–91:

"The significance of the variation of the value of money is taught by history. While our theoretical experts keep harping on the necessity of a return to sound money (by which they mean the money of yesterday), history shows that for centuries and centuries, in fact ever since the invention of money, money has never ceased to depreciate. The point, therefore, is not to restore the value of money, but rather to devaluate it methodically. . . . The fall of the value of money has indeed invariably manifested itself as a concomitant of an increase of business. Is it revolution to take account of an evolution of such old duration? . . . The point at issue is to provide against an unregulated and haphazard depreciation and to obtain from depreciation, *by regulating it*, the maximum effect of which it is capable."

This author, then, advocates inflation as a permanent institution, while he has no remedy to offer against the evils resulting from a progressive depreciation of money. I have recently read an utterance of M. Charles Gide's, surely one of the first French authorities, to the effect that the steady decline of the price of gold down the ages has been the motive force of economic progress, and that the only method for overcoming the unreliability of the money machine is to introduce an element of constancy into the rate of depreciation. Here we have the idea of Silvio Gesell's *Schwundgeld*. Waning money it must be. But a depreciating currency must be guarded against at all costs. The great thing is to discover a method for combining depreciating money with a stable currency. All that is needed to find the right proceeding is to understand that interest can act as a depreciator of money and claims of money, and that interest, at the same time, can act as the main stabilizer of the currency.

a house, a piece of land, a machine, stocks and shares; for interest, and some future purchase, if he invests the money in the savings bank; for a reserved seat in Heaven, if he gives it to the Church, or to the poor that are always with us. Gesell himself defines money as effective demand unqualified. Now money, which is so constituted as not to suffer demand to be deferred one day, must needs render demand more ready, more urgent, more powerful. Gesell performed a kind of logical somersault in the train of his argument when he deducted from the money reform, as its main and initial development, an increase of production and supply. He overlooked the anterior stimulus to consumption and demand. He took the second step before the first. Consumptive money can have one effect only: to stimulate consumption. Only as a consequence of this, in the wake of stimulated demand, can a livelier tone begin to quicken the pulse of the economic organism, from which an increase of output will originate. Consumption must set the pace; it is the primary element, the only true motive of production.

Now interest is the expression of the ratio between supply and demand: Gesell's own definition, and the correct one. The more demand exceeds, and precedes, supply, the greater must interest be. Consumptive money, this incessantly threatened and harassed money, is all effective demand; it pants for goods and real capital; it strives to escape out of itself: it stimulates demand, so that demand is strengthened and eternally ahead of supply, exceeding supply. Gesell's—and Professor Soddy's—conclusions are based on the assumption that the effects of a more active money would first begin to bear on the other factor, output and supply, which is undoubtedly wrong.

From these considerations it follows that Gesell's vision of money accumulating in private hoards, if not made subject to material shrinkage, is an illusion. It implies that people are enabled to acquire real capital, and to produce it in unsuspected profusion, while retaining the surplus of their earnings in the shape of money—eating their cake and having it. Production is the most powerful consumer. It consumes right and left: on the one hand raw material, on the other hand old capital. A new machine scraps an old one, which ceases to be capital; automobiles diminish the usefulness of railways, which lose some of their capital virtues. Capital can only come out of capital, and as much as is born, just so much is destroyed. Money, which is the reflection of goods, shrinks away with goods; to inflict on it an extra and artificial shrinkage would therefore be to expose goods and real capital to an extra and artificial wastage.

Far from causing interest to dwindle and vanish, consumptive money would over-stimulate interest and consumption. Any form of stabilized money will rather favour than oppress interest, because it encourages consumption. Indeed, the essential condition of a successful stable currency is well regulated, unfailing, prompt consumption, with demand always exceeding and preceding supply. Thus the money reform assures the existence of interest; it extricates interest from the contracting forces which hitherto have so frequently hampered the display of its energies.

Does this mean that the vested interests are safe? Does it lead to the desolatingly hopeless conclusion which Mr. Keynes has taught us to envisage? The permanence of interest, not only simple but compound, with the concomitant periodic upheavals of the social and economic foundations? It cannot be. The stability of the purchasing power of money is an idea so thoroughly just and natural; the general trend of conditions points so unmistakably and irresistibly in the direction of its recognition and realization, that it is inconceivable it should perpetuate and even aggravate the evils of interest. It cannot but alleviate and eliminate them. Interest in itself is not an evil and not a social wrong. The condemnation of interest by sage and saint all down the ages has been a misconception. Interest has been blamed for the misdeeds of another. The undeniable evils—so clearly and forcibly set forth by Mr. Keynes, so indignantly denounced by Professor Soddy—which interest has seemed to bring down upon civilization are entirely due to the fluctuations of the currency standard. These fluctuations have been making of interest the instrument of exploitation : now exploitation of creditors by debtors (in periods of money depreciation), now exploitation of debtors by creditors (in periods of deflation ; the wage workers, in so far as they are not owners of real estate, are to be considered as debtors, since they work with means of production not owned by them). When the fluctuations of the value of money cease, interest will become a fixed and constant quantity, appearing both on the credit and the debit page of the workers' account-sheets. Workers will charge the interest they pay to their wages, which they can safely and successfully do, because their services are sought after ; entrepreneurs and merchants charge it, as they have always done, to their prices ; the professional politicians to their party funds and bribes ; the State to its taxes. In other words : interest is not really paid over the counters of banks, but in the innumerable small and great daily payments. Interest-gatherers are all workers by hand or brain, all earners of working incomes, masters and men alike ; payers of interest are all consumers. This amounts to saying that, in the last resort, interest is paid by those who are consumers without being earners, and that, consequently, it is interest, the supposed usurer and exploiter, which eats up the vested interests and "arms enterprise against accumulation." Unearned income has no chance of charging its payments of interest on to others ; but it has to bear, in providing for the cost of living, the extra charge of prices swelled by the element of interest. It is not all too difficult to demonstrate or comprehend that with a perfectly and lastingly stable currency, in spite of an invariable rate of interest, the interest bearing investments must be gradually dissolved, consumed.

It is an observed fact that interest-bearing savings are formed in periods of prosperous trade. When prosperity is uninterrupted these savings will assume much handsomer proportions yet. But they cannot be formed, unless old fortunes and savings are consumed. For, as in all nature, new births have to be induced and counterbalanced—I would say, paid for—by deaths. A nation, or in fact the world of nations, cannot economize on its output. Saving means to produce more than what is consumed; for individuals to be able to save, there must be other individuals who consume more than they produce. Here we touch upon the process which reduces vested fortunes, big and small.

These considerations—a few more will be added farther on—should allay the misgivings which Mr. Keynes raises in his reader and some of Professor Soddy's indignation. Only in periods of stagnation, when its weight is doubled and trebled by the fall of prices, does interest become a burden and a menace. The stable currency standard, obtained through stability of the rate of discount, will make an end of stagnation and so an end of usury.

II

WHAT IS A STANDARD OF CURRENCY?

A. The Gold Standard

A STANDARD is what stands, what stays, what holds its own and endures. The gold standard theory, true to this definition of the term, rests on the principle that a given weight of gold shall always be bought and sold by the Central Bank for a fixed amount of legal tender. This method of dealing in gold (not *with* gold) has given rise to the notion that gold has immutable intrinsic value. Some critics have scornfully paralleled this idea with the wildest superstitions of the Dark Ages. Experience indeed seems to have proved that the theory is a fallacy. The price of gold is not an absolute quantity, neither is its value; it is made and incessantly remade as a ratio to the price of other goods, that is to say, it can only be expressed in terms of these other prices. Also the degree in which gold is esteemed—valued—by men varies greatly. However, that is not all that may be said of the price and value of gold. I have already shown that with a stable currency the price of gold must be stability itself, and I have indicated the reason why: namely, the chemical constitution of the metal. Here I must attempt once more to explain why it was absurd to fix the price of a weight unit of gold at a definite number of monetary units. It was absurd so long as the value of this monetary unit was not strictly defined and fixed. The gold standard theory, expressed in terms of an equation, would be to this effect:

One ounce of sterling gold is equal to $3x$ $17y$ $10z$.

It is an unknown quantity. For, what does dollar, or pound sterling, or franc, or mark stand for ? In what shape do they present themselves ? Either as coined gold, or as silver, or as paper money. If as coined gold : to say that a given weight of gold equals a fixed sum of money is to say that uncoined and weighed gold is bought with coined and weighed gold. If as silver or paper money : to say that a given weight of gold corresponds to a fixed sum is to say that the price of gold *is determined by* the price, or value, of silver or paper money. How could responsible men, in the age of exact science, continue to constitute the material of the thing to be measured as the standard of measurement ? Gold standard signifies that gold is to be the measure of money, which money is not itself the thing to be measured. But gold is not only the measure of money, it *is* money. Thus the term of gold standard—considered to be scientific—is found to signify that money is the measure of money (gold the measure of gold). It leaves the thing to be measured without any measure at all. Gold standard as a measure of value is as ingenious a device as would be the use of one part of the body as a thermometer to take the temperature of another part. The gold standard theory was not a theory of buying and selling and of price-formation, but a theory of money-changing. To-day the waking minds have come to understand that the question of a monetary standard is not one of the ratio between various kinds of money, but one of the ratio of money (no matter in what material form) to that which is bought for money and sold for money : goods. And the conclusion imposed on their reasoning by this new insight is that a standard should ensure the stability of price.

For a hundred years and more the problem of a standard of price has from time to time taken hold of some intelligent brains. Silvio Gesell is entitled to the fame of having taken it up in the nick of time (when everybody had abandoned it), but more especially of having been the first to grasp and demonstrate its full significance. He began as far back as 1891, and he worked at it twenty years. His contribution will come to take first rank.

B. The Index-Number Standard

The gold standard in its traditional make-up is doomed to go, for it has failed. Mr. Keynes has found the apt word for it; he calls it "a barbaric relic." What will come in its place? So far the Index-number standard is the only alternative that has been suggested and elaborated.[1] But it presents insurmountable difficulties. It took Mr. Keynes a long time to convert himself to it; we shall see, moreover, that he has, even now, not overcome all his reserves. (I am inclined to believe that it was his heart, rather than his head, which vanquished Mr. Keynes. He is one of the bright hopes of our time through his noble passion for the redress of old wrongs and the healing of new wounds.) Professor Gustav Cassel of Stockholm University, an approved authority, has been hovering about the threshold of the new temple; he is deterred from crossing it by practical, or tactical, considerations. At the bottom of them, I doubt not, there is a residue of distrust in its theoretical soundness. As to the men in charge of the currencies, the famous Havensteins and Dr. Schachts, they are too busy, and too self-satisfied, to try and familiarize themselves with new theories.

But how is it that an innovation so obvious and so promising as that of a stable standard of price, of "the value of money," should meet with so much opposition, distrust, indifference? People could not be made interested

[1] Mr. Lowenfeld claims that the scheme proposed by him would insure stability. In making this claim he takes for granted what he has forgotten to prove. The German writers who elaborated the theory from which the scheme is derived never pretended that it would make for stability; indeed, they did not so much as conceive the idea of stability.

in aviation or wireless telegraphy until the possibility of the thing was demonstrated to them by successful experiments and performances. Now currencies are not a thing to be experimented with (though, surely, they have been managed in a darkness much denser than that enwrapping the first steps of adventurous alchemists). In the place of the successful experiment we need an imposing body of compact and perfectly consistent, unified scientific and expert opinion, first as to the practicability of the idea, and second as to the ways and means of its execution. We need not despair of this. More than half the battle has been won, since the nations have come to realize that the stability of price is an essential condition of economic, political, and social welfare and peace.

The notion of Index-number standard is as yet far from being uniformly understood by its advocates. There is something elusive about it, which baffles students. Its principle consists in this : that the movements of the level of prices are to be made use of to control this same level of prices ; the measures employed to stabilize the level are to be chosen according to its fluctuations. Does not that smack of gold standard ingeniosity ? Is it not fundamentally the same conception as gold standard theory ? Measure the measure by the measure and declare that it comes out pat ? Is computation the same as measuring—provided a monetary standard can at all pass as a measure ?

If I want to keep a thing fixed I have to place myself outside of it. I have to start from what is not the thing to be managed. In order to regulate the flow of prices we must not ride on its current, but stand at the head and control the forces that govern price. The usual objections to the Index-number standard do not touch upon this essential point. They aim at the contingent parts of the scheme : that the statistical computation of the Index-number cannot be made accurately enough or promptly enough to give satisfactory results; that the so-called credit factor is uncomputable and unmanageable, etc. But these superficial and vague objections would be speedily enough overthrown and discarded if it were not for an underlying indefinable feeling that the essentials are untenable.

The fundamental vice of the scheme seems to me to consist in the fact that this general level of prices, which is to be the standard of value, does not represent any unit at all. What should we think of a linear standard measure composed of a hundred particles of a hundred different materials? The level of prices, by whatever methods it may be computed, can always be made the subject of contention. What has to be recomputed month by month surely cannot claim to pass muster as a constant standard, removed from all chances of miscalculation or arbitrary interference. Gesell and Professor Soddy are satisfied that the Index-number computation suffices for a perfectly reliable management of the money circulation. (Gesell is justified in doing so because he relies on the virtues of consumptive money. I have already expressed my opinion that a regular periodic depreciation of the money tokens would ensure perfect regularity of their speed of circulation and thereby the stability of the price level). Also Mr. Goldsborough's Bill does not make any further provisions. As to Professor Fisher, he is represented by Mr. Keynes as holding the same view; but in his more recent article he has come round to Mr. Keynes's own more sceptical standpoint. He says: "Since the gold base and the paper and credit superstructure are thus parts of one total circulating medium, both should be regulated. To regulate either alone, though it would tend to regulate the other, would not always be sufficient. . . . The flow of credit is therefore governed by the rate of interest." So Professor Fisher abandons his single standard to adopt a double one, which is a great deviation from sound scientific principle. Mr. Keynes goes a great deal farther. Here is a passage showing what he considers necessary for the successful control of the price-level (p. 188–9):

> Actual price-movements must, of course, provide the most important datum; but the state of employment, the volume of production, the effective demand for credit as felt by the banks, the rate of interest on investments of various types, the volume of new issues, the flow of cash into circulation, the statistics of foreign trade and the level of the exchanges must all be taken into account. The main point is that the objective of the authorities, pursued with such means as are at their command, should be the stability of prices.

Where we expected to be supplied with a real standard, fixed, definite, undisputed, and undisputable, we are thrown upon the judgment and discretional action of "authorities," which means of hired men, hirelings. Where we are entitled to proofs, we are asked to believe, and to believe in the judgment and discretion—bearing on a multiplicity of factors—of officialdom. It will not do.

Generally speaking the Index-number standard is not a measure to go by, but rather a result to be aimed at and, if possible, arrived at. There is no commanding principle, nothing that is really ascertained, settled and agreed upon. The observed facts, from which Mr. Keynes's arguments are drawn, are admirably chosen and interpreted; but somehow they do not seem to go to the bottom of the problem. Index-number can no more be recognized as a standard than an abstract number can pass as a definite quantity. There are no terms in any language to express it. For, surely, it would be a mockery to say that the standard of price is 175, or 150, or 425; these figures being supposed to express the standard cost of living at a given date.

All science strives after unity, after the all-inclusive one, and does not rest content until it has won through to the core. There must be, underlying the multiplicity of phenomena, a fundamental and central fact in which all the various manifestations are focused into an indisputable and clear unity governing all. It must be possible to discover a standard of price which is not a compound but a unit, not a result but a principle, and which, moreover, is not itself a ware or the price of a ware. What is it? It is the thing which, of all others, Mr. Keynes has overlooked in his enumeration: consumption. We shall be on the road to a solution when it is recognized that consumption is the basic factor in economic life, the originator of interest, and therefore the actuator of all the economic forces; when it is understood that interest is the visible manifestation of the economic energy, and that all the derived factors, such as prices, wages, production, credit, etc., can be controlled through it.

C. The Interest Standard of Currency

Interest is the natural and concentrated expression of the ratio of supply and demand, which determines price. And it would be a serious error to think that interest only applies to so-called capital and not to goods also. A thing is capital only in so far as it is a ware, having a price; and in so far as it has a price, a ware (and likewise the commodity called labour) is capital. A price is formed wherever interest manifests itself, so that a person is induced to demand, inquire, about the price of a thing. The price will be higher or lower in proportion as the interest manifested is more or less lively and genuine. The readiness to pay interest is the reliable test of the need for the article. Now, the need for a thing begins to manifest itself when things have disappeared and the thing itself is disappearing, or threatens to disappear, through consumption. Demand for wares arises, not when they have been produced, but when they are being snatched up. Production follows consumption. It is the process of replacing what consumption has taken away. I use the term "consumption" in its broadest sense. Nietzsche says: "A need is considered as the cause of things coming into existence, but more truly the need arises because things have come into existence." That is very aptly conceived. No sooner has man created a thing than he feels a new need created in him. Most of our needs spring from our possessions, from the things with which we surround and overwhelm ourselves. For the making of things annihilates or consumes things; production, I have already said, is the great consumer. New inventions consume the readiness of men to put up with antiquated conditions. From this destruction, this displacement, this consumption; from this never-flagging impatience with antiquated conditions; from this ineradicable and insatiable human appetite does interest—the word taken in its two senses—arise. Where there is no consumption, there is no interest. Interest is the measure, the meter of consumption, of demand. Hence interest is also the measure of the demand for money, since demand cannot become effective except through the supply of money. If this argument is correct, the theory of an Interest standard of price presents itself spontaneously. Here it is summed up in a few propositions:

(1) Interest, and supply of money stand in a constant ratio to each other.

Men are willing to pay the price—supply the money—provided their interest is sufficiently strong. Supply of money is a combination of two factors: energy (the velocity, urgency of circulation), and matter (the volume of money issued). But the latter will constitute itself automatically if the former is given; which it is when the rate of interest is fixed. There are no statistics needed to determine the volume of currency required.

(2) If one of the two factors changes, the other must needs change at the same time; rising prices go with rising interest, falling interest brings down prices.

The order is usually this: a waxing or waning interest

in goods, as manifested by men, calls for a rise or decline of prices; the rate of interest—interest in its concreter shape—follows suit. It is the tripartite order universally observed: interest—price—interest in another shape. Aristoteles' idea of becoming (*fieri*) and being (*esse*) is to this effect: actual being comes from potential being out of an actual being; the order is the same as above: actual being—potential being—actual being of another kind. In the philosophy of Schopenhauer we get another very apt example of it in the relation of spiritual and material: first is will (spiritual), next the body with its brains (material), last intellect (spiritual, but in another form).

(3) If one of the factors is stabilized, the other will stabilize itself also. An unalterable interest creates an unalterable demand for wares and supply of money.

(4) The aim of the reform is the constancy of the general level of prices, which level is the expression of the ratio between supply of money and supply of wares.

(5) A ware is what is offered for sale against money. Wares are produced because it is hoped that they can be exchanged for money. Wares are made according to price, not vice versa. Hence the supply of money as governed by interest (men's interest in wares), governs and creates the supply (production) of wares. Production follows consumption, which creates interest (second stage).

(6) By putting the supply of money under a constant pressure of unalterable magnitude we obtain an unalterable ratio between the supply of money or demand for wares (interest in wares, readiness to pay the price to which the interest element is added), and the supply of wares (demand for money, wares having been produced with a view to being sold for money): from this stable ratio results the stable level of prices.

(7) The pressure on the supply of money is brought about by consumption; the need and interest arising from consumption, and the measure of the pressure (the

urgency of the need and interest) is expressed in the interest conceded.

(8) Consequently a constant pressure (rate) of interest must in the first place produce a constant supply of money (which is expressed in a constant velocity of circulation; the volume of money issued adapts itself automatically, as pointed out above); in other words, the regular disposal and consumption of the wares produced, which in its turn induces an equally regular supply of wares. Out of the steadiness of the ratio between the two factors results the steady, unalterable general level of prices, the stability of the purchasing power of money; in a word: the true standard of price. The rate of interest is revealed to us as the natural controller of the currency.

III

FLUCTUATIONS OF INTEREST AND FLUCTUATIONS OF PRICE

The suggestion of fixing the rate of interest once and for all will have much scorn poured upon it. To forestall unnecessary scepticism I would beg the reader to understand that what I propose to fix is not the rate of interest proper, but only the official rate of discount. It stands to reason, however, that a fixed rate of discount will stabilize the average rate of interest. Interest has always been considered as the freest and least controllable of things. Attempts, how often repeated, to fix it within limits have signally failed. And interest, moreover, is looked upon as the safety valve of the economic machine, or as the barometer and thermometer of the economic atmosphere. To nail it to the plank is to deprive ourselves of our most trusty guide and safeguard. All these considerations are valid so long as we have no standard of price, so long as interest is forced to keep moving in order to prevent the economic equilibrium from being upset by the vagaries of other forces. But they are fallacious as soon as the mechanism is more accurately understood, and interest is assigned its legitimate rôle and function.

In the first place we must learn to look upon the rate of interest as a mere screen to hide reality. When the rate goes up, the real value of interest goes down, and vice versa. The fluctuations of the rate of interest " are not what they seem." Especially the fluctuations will be found to have been much smaller than they appeared to be, if we but take into account certain extraneous elements that will dress themselves up in the uniform of interest. Interest is to be compared to a musical string well fixed to its two pegs. When pressed it will vibrate. But to vibrate means to stay fixed and to continue in the original station. The stability of the rate of interest is no less the normal condition of the economic body than the constancy of temperature is the normal and healthy condition of the animal organism. We should learn to recognize its aberrations as a sign of disease, and be warned that something has gone wrong. The general diagnosis will be given below.

How little even the experts have understood the mechanism of currency and interest is forcibly brought out by certain contradictions in the schemes of the reformers. Says Mr. Keynes (p. 21):

> It is for this reason, amongst others, that a high bank-rate should be associated with a period of rising prices, and a low bank-rate with a period of falling prices. The apparent abnormality of the money rate of interest at such times is merely the other side of the attempt of the real rate of interest to steady itself.

Let us fix in our minds the fact observed and the reason for it given by Mr. Keynes: *That the rate of interest* (*money rate*) *and the general level of prices* (*money prices*) *move on parallel lines*. And do not let us pass lightly over Mr. Keynes's opinion that there are more reasons than one why this should be so. I call this a very momentous discovery; the author himself seems rather amazed at it and half-incredulous. There is, perhaps, no stronger proof of the value of Silvio Gesell's contribution to the science of money than the fact that he discovered this parallelism twenty years before Mr. Keynes. Here is a passage from *Aktive Währungspolitik*:

> The records of the rate of interest prove that the rate of interest has risen whenever the circulation of money has been increased. It

has made no difference whether such increase was brought about through increased coinings, or increased issues of bank-notes, or dilution of the metal, or increasing speed of the circulation. The rate of interest rose to an unaccustomed point when the bands of Pizarro flooded Europe with gold; it rose when the Californian gold finds were minted; it rose in Germany when the country revelled in the French war indemnity of 1871. . . . When John Law tried to satisfy the merchants' thirst for money with paper money, the rate of interest rose; when, during the French Revolution, the attempt was made to "coin" the soil in the shape of assignats, the rate of interest rose; and it rose when the Reichsbank tried to reduce it through increasing issues of bank-notes. . . . And also the other way about. The rate of interest fell in the Roman Empire when the supplies of gold from Africa gave out and the silver mines of Spain were exhausted; it fell when the treasures of Montezuma and of the Californian mines were exhausted; it fell in Germany when the boom of the indemnity was succeeded by the slump in consequence of the export of gold.

Gesell understood and knew. However, from his insight and knowledge he drew conclusions and made practical suggestions which contradict pretty nearly everything that is implied in the facts. I have already dealt with his contention that under a stable standard not safeguarded by shrinking money the rate of interest would fall and an excess of currency would be issued. How could that be possible if it is proved that an increase of the volume of currency forces up the rate of interest? No contradiction could be more complete. And it is not a casual one, an inadvertence; Gesell has repeated over and over again the arguments which constitute it, and he has ridiculed those who seemed ignorant of the facts taken singly. Nothing can shake my faith in the superior genius of Silvio Gesell, but his errors make me realize how desperately entangled the strands of monetary theory must have been when he took up the study of it.[1]

In the passage quoted the argument proceeds from money to interest: the changing volume of circulation forces the

[1] *C'est le privilège du vrai génie, et surtout du génie qui ouvre une carrière, de faire impunément de grandes fautes.*—VOLTAIRE.

rate. Not till 1923 did Gesell come to state that the rate of interest might determine the circulation of money and the level of prices. In an article published in August 1923, he explicitly says that a high rate accelerates circulation, while a low rate impedes it. But even now he has not disentangled himself from the old error; for he has not revised his theory of discount, which is the accepted one and to the effect that, in order to speed up the supply of money, the rate of discount should be lowered. Where he sets forth his proposals as to the regulation of the issue of notes he points out that the issue can be augmented by making borrowing easy and attractive through a low rate of discount.[1] Along with the true recognition of the natural facts, Gesell, very inconsistently, retains the belief in the traditional banking theory. And so does Mr. Keynes, as I shall now demonstrate.

After having established the fact that, for more reasons than one, high prices should be associated with a high bank-rate, he proceeds in all his practical suggestions from the opposite conception. Consider for instance his treatment of the "Quantity Theory." Mr. Keynes constructs his argument on rather novel, and very ingenious, lines. From Gesell's and Professor Fisher's method he differs chiefly by introducing into his equation the term of "consumption unit," which is Gesell's "velocity of circulation" turned into sloth of circulation; for the consumption unit stands for the quantity of money retained by the public, i.e. not circulated. Before I can quote the passage relating to the influence of the rate of discount it seems necessary to reproduce Mr. Keynes's demonstration (p. 77):

Let us call such a unit a consumption unit, and assume that the public require to hold an amount of money having a purchasing power over k consumption units. Let there be n currency notes or other forms of cash in circulation with the public, and let p be

[1] See *Aktive Währungspolitik*, p. 57; the passage is quoted below.

the price of each consumption unit (i.e. p is the Index-number of the cost of living), then it follows from the above that $n = pk$. This is the famous Quantity Theory of Money. So long as k remains unchanged, n and p rise and fall together; that is to say, the greater or the fewer the number of currency notes, the higher or the lower is the price level in the same proportion.

The formula of the theory has to be extended to include the influence of "credit."

Let us assume that the public, including the business world, find it convenient to keep the equivalent of k consumption units in cash and a further k' available at their banks against cheques, and that the banks keep in cash a proportion r of their potential liabilities (k') to the public. Our equation then becomes

$$n = p\,(k + rk').$$

So long as k, k', and r remain unchanged, we have the same result as before, namely, that n and p rise and fall together.

Now for the discount policy (p. 85):

The usual method of exercising a stabilizing influence over k and k', especially over k', is that of bank-rate. A tendency of k' to increase may be somewhat counteracted by lowering the bank-rate, because easy lending diminishes the advantage of keeping a margin for contingencies in cash. Cheap money also operates to counterbalance an increase of k', because, by encouraging borrowing from the banks, it prevents r from increasing, or causes r to diminish. But it is doubtful whether bank-rate by itself is always a powerful enough instrument, and, if we are to achieve stability, we must be prepared to vary n and r on occasion.

The outstanding question raised in this passage is: when does money (cash) accumulate at the banks—or in private hands? First of all let us agree that it makes no difference where it does collect, whether in stockings, mattresses, and safe-deposits (k), or in the banks (k'). Cash which the public store away is cash that finds no employment. When people take their money to the bank, it is because they shrink from venturing it in enterprises and purchases. This is a sufficient reason why cash balances at the banks are swelled in periods of depression and unemployment. Money is affected in the same way as labour : it is forced to lie idle when prices flag.

Now, what is Mr. Keynes's advice as to the bank-rate under these circumstances ? He says : lower it. And what was his opinion in the passage quoted ? That a high rate should be associated with a period of rising prices. And what is the object of the discount policy ? To prevent prices from falling when they tend to fall, from rising when they tend to rise. In the situation under consideration all the symptoms spell slump : falling prices, slow-moving money. And again I ask : what is the expert's advice ? Lower the bank-rate ! This, then, is the argument :

(*a*) Falling prices are associated with a low bank-rate. The logical conclusion to be drawn from this is that the lowering of one (the rate of discount) must lower the other (the level of prices), and that the fall of prices can be counteracted by not allowing the rate of discount to budge ; but the conclusion drawn by the experts is the reverse of this :

(*b*) Prices are falling ; in order to counteract this movement it is necessary to lower the bank-rate.

Apply the argument to a more familiar phenomenon :

(*a*) A failing light in the kerosene lamp is associated with the lowering of the wick.

(*b*) The light is failing ; in order to counteract this movement it is necessary to—screw down the wick.[1]

[1] I quote from an article by Mr. Keynes, which was published in the *New Leader*, June 20, 1924 : " Any sensible person is in favour of dear money (by which he means a high rate of discount) in certain circumstances and cheap money in other circumstances. In the ordinary course the present slump will come to its final conclusion, and, just as on previous occasions, a boom will begin. If by the date at which one could hope that some of these ideas may be beginning to fructify the slump is thoroughly over and the boom beginning, we shall make our first experiment in the form of putting on dear money at a very early date compared with previous occasions, and avoiding the impending boom of 1925." That is to say, when prices begin to look up we shall raise the bank-rate.

Experts have often wondered that the handling of bank-rate has not worked according to expectations. In 1923 an article by Professor Gustav Cassel, who is considered a great authority on questions of currency, was widely circulated, dealing with the probable effects of the raising of the bank-rate in England which had taken place a few weeks previously. After discussing the pros and cons of the case the expert wound up his argument by saying: it will be interesting to watch events; the level of prices may go up, or it may go down; also it may stick. "Good Launcelot Gobbo" was in much the same predicament of uncertainty when he was trying to make up his mind about leaving Shylock the Jew. It is to be noted that Professor Cassel had advised the Bank to raise the rate, the purpose of the measure being to improve the exchange-rate of the sterling, i.e. to bring down prices, enhance the the value of the currency. And it is to be specially noted that in due course of time the English Index-number began to move upward.

Such is the point at which expert opinion stands to-day with regard to this vital problem. Professor Fisher shares the prevalent (and undisputed) notion. He says: "The rate of interest is like a valve. Tighten it and less credit will flow. Relax it and more credit will flow." The school of Silvio Gesell staunchly stick to the dogma, the master abetting them. Reduce the rate of discount, was the burden of all the proclamations of the Swiss money reformers in combating the policy of the Bank of Issue since the outbreak of the crisis in 1920. The Bank did reduce the rate and matters grew worse. But nobody seemed to heed results and be any the wiser. The Swiss currency reformers continue to give advice to this day. In May 1924 their mouthpiece raised a warning against the coming on of a new depression: "We are drifting towards a new crisis and a recrudescence of unemployment unless the National Bank immediately reduce the bank-rate."

Mr. Keynes, who surely has been a very attentive and penetrating observer, has made some progress in so far as he has become sceptical regarding the efficacy of bank-rate policy. And he goes so far as to speak of "a mistaken bank-rate policy" (p. 136). The passage deserves to be quoted in full :

> Dear money—that is to say, high interest rates for short-period loans—has two effects. The one is indirect and gradual—namely, in diminishing the volume of credit quoted by the banks. This effect is much the same now as it always was. It is desirable to produce it when prices are rising and business is trying to expand faster than the supply of real capital and effective demand can permit in the long run. It is undesirable when prices are falling and trade is depressed.

What is "dear money" ? Is it only the rate of discount that makes it dear ? Has Mr. Keynes forgotten what he wrote on page 20 ?

> Economists draw an instructive distinction between what are termed the "money" rate of interest and the "real" rate of interest. If a sum of money worth 100 in terms of commodities at the time when the loan is made is lent for a year at 5 per cent. interest, and is only worth 90 in terms of commodities at the end of the year, the lender receives back, including his interest, what is only worth 94½. This is expressed by saying that while the money rate of interest was 5 per cent., the real rate of interest had actually been negative and equal to minus 5½ per cent. In the same way, if at the end of the year the value of money had risen and the capital sum lent had come to be worth 110 in terms of commodities, while the money rate of interest would still be 5 per cent. the real rate of interest would have been 15½ per cent.

In consequence of the raising of the bank-rate the price-level went up, which amounted for the borrower to a debtor's premium at the expense of the lender. So then it is not the raising of the discount-rate that makes dear money and scares off borrowers—in fact it rather attracts them, and naturally too, because it makes money cheap, extra cheap.

When might-be borrowers fight shy of borrowing, it is from causes widely different. They are deterred by official proclamations, by the auguries of Sir Such-a-One, that the level of prices will be consciously and deliberately brought down. Such a proclamation, loudly acclaimed by all the authorities and most of the experts, was solemnly issued by the International Financial Conference at Brussels in 1920. There is nothing more masterly in Mr. Keynes's book than the page setting forth the natural effects of a deflationist measure thus agreed upon and proclaimed (p. 144).

> The policy of gradually raising the value of a country's money to (say) 100 per cent. above its present value in terms of goods amounts to giving notice to every merchant and every manufacturer that for some time to come his stock and his raw materials will steadily depreciate on his hands, and to every one who finances his business with borrowed money that he will, sooner or later, lose 100 per cent. on his liabilities (since he will have to pay back in terms of commodities twice as much as he has borrowed). Modern business, being carried on largely with borrowed money, must necessarily be brought to a standstill by such a process. It will be to the interest of every one in business to go out of business for the time being; and of every one who is contemplating expenditure to postpone his orders so long as he can. The wise man will be he who turns his assets into cash, withdraws from the risks and the exertions of activity, and awaits in country retirement the steady appreciation promised him in the value of his cash. A probable expectation of deflation is bad enough; a certain expectation is disastrous. For the mechanism of the modern business world is even less adapted to fluctuations in the value of money upwards than it is to fluctuations downwards.

This is the debtor's case. Mr. Keynes might have mentioned, in this connection, the anxiety of creditors about their "frozen credits," the reluctance of capitalists to venture their available surplus in lending it to needy borrowers, the many bankruptcies, the losses of invested capital, etc. By looking at the situation from the creditor's point of view we best learn to understand why capitalists, i.e. those who have money to lend, are not in a hurry to try and hook the 100 per cent. which the borrowers lose: it is all too risky a stroke, because it is likely that the angler will be hauled down and disappear under the ice—rod, line, hook, and all. Is it to be wondered at that money (cash) should be lying idle, a mighty abundance?

In the face of such a situation so created by a declared policy of deflation the declaration that the bank-rate has been raised must have the effect of a warning in all the ranks of business—with the happy exception of the bearish speculators, whose path to glory is beautifully cleared and smoothed. Both would-be borrowers and lenders are paralysed. If it were not for the proclamation and the various trumpetings flourished out to work upon the expectations of the public, the raising of the rate of discount would surely cause a rise of the level of prices. A little bit of observation of actual occurrences will corroborate my contention. The instance of England in 1924 has already been adduced. In Switzerland exactly the same happened: the bank-rate was raised because the rate of the franc had been declining, and it was expected that the tightening of the credit valve would arrest the upward move of prices. Nothing of the kind, but prices kept advancing, credit expanded cheerfully, and business looked up. The most notable instance of the raising of bank-rate being followed by expansion of credit and prices was the time after the outbreak of the war. All the Banks of Issue the world over raised the rate of discount—with the avowed purpose of protecting their reserves, i.e. their credit—and, sure enough, in all countries the world over, a few weeks later, prices began to move upwards. Not in spite of, but because of bank-rate policy.[1]

Thus Mr. Keynes affects to criticize the Bank of Issue

[1] The effect of a forced raising of the bank-rate at the end and turning-point of a period of inflation must needs be different from the effect produced by a raise at the beginning of a boom.

while simply contradicting his own truer insight. The Bank does practise the method which he justifies and advocates; it raises the discount-rate when it wants to put on the brakes, and it relaxes the valve when it wants to make things go full steam. It does what the expert pronounces to be desirable. He can only criticize it for not acting promptly enough and thoroughly enough.

Merely to show how general the belief in the discount fallacy is, I will mention what Mr. Brailsford, the editor of the *New Leader*, and one of the best heads among leading British Socialists, writes in his review of Mr. Keynes's book. It is to this effect: the men who at this time are probably the ablest experts in matters of finance in England believe that by such a simple measure as the reduction of the bank-rate to 2 per cent. and the discounting of some ten million Treasury Bills unemployment might be overcome. Like Mr. Keynes, Mr. Brailsford holds that the main cause of unemployment is the fall of prices. In order to overcome unemployment it would be well to stop the fall. In fact the *New Leader* had been vigorously advising that a dose of inflation would be the remedy.

Now let us consider what the effect on prices the lowering of bank-rate must naturally have. In the same way as the fall of prices, far from attracting buyers, deters the would-be buyers, the fall of the discount-rate must produce a deterrent effect. A house built with "cheaper money" will be a cheaper house. The cheapening of the money, like the cheapening of coal and other basic materials, cheapens everything else—in the course of time. *For cheapening is a progressive process:* six months hence things will be cheaper than they are now. Therefore he who starts building his house six months hence will get a cheaper house than the fool who rushes in to-day. And there are fewer fools than we are apt to believe. The lowering of bank-rate does not start houses building, much less factories; but rather the reverse.

Both practical and theoretical experts have been overlooking the fact that the stimulus for the withdrawal of money is not given by the discount policy. But it is time they should become aware that more often than not the raising of the bank-rate is followed by a rise of prices. These experts further overlook the fact that business borrows in two places : at the banks and at the Bank of Issue. Therefore if the latter, in order to attract borrowers, reduces its rate, the immediate effect will be that nobody wishes to borrow at the banks, which have a higher rate. Instead of new loans being sought, the chances are that the old ones will be cancelled by the debtors. Even if it were possible to reduce the interest on all loans, including those issued at a fixed rate, the effect would be disastrous : namely, a fall of prices even more general. Creditors would be made happy to get their credits liberated. The moving of bank-rate downwards creates a general disturbance. It relaxes every tension and so paralyses the economic forces. Far from stimulating business and accelerating the circulation of money, the lowering of bank-rate must normally drive money out of its active employments into safe places and bank deposits. Professor Fisher's recipe : "tighten the valve and less credit will flow ; relax it and more credit will flow," is an epitome of inverted policy. To relax is to lessen the tension. However, it is tension that matters. If you want a display of the forces of credit (in which cash is included) you must sharpen the edge of desire in borrowers (consumers of credit). The contrary effect is produced when they are specially asked to please come and help themselves ; all the more so when they are warned that they will be made to disgorge through price-cutting.

Although he does seem to have got stuck on the mud banks of his theory of interest, Silvio Gesell has lately come to the rescue of my contention. In one of his more recent publications (May 1924) he writes :

A high rate of interest has very often been pronounced to be the cause of money shortage. With all that we know to-day that the rate of interest must go up along with prices, and that prices rise when too much money has been issued, so that to suit the issue of money to the requirements of the rate of interest would only force up the rate of interest all the more.

Gesell contends that a high rate of interest is not the cause of money shortage, but rather tends to produce an excessive issue of money. And nothing could be more obvious than that it should be so, since an exorbitant rate of interest is bearable only if counter-balanced by high prices : the debtor's disadvantage is made up for by the debtor's advantage ; the creditor's advantage is cancelled by the creditor's disadvantage.

Nothing is more amply proved than the fact that the rate of interest (though not by any means the real value of interest) falls and rises with the level of prices. We shall have to learn yet that it is the rate of interest which governs the level of prices. The observed facts all go to prove that the principle first established by Mr. Keynes, but persistently repudiated by him subsequently—to wit that rate of interest and level of prices move on parallel lines—is a natural law and must have a natural cause. This law, rightly considered, should suggest this first and main interpretation that, since the level of prices moves up and down with the rate of interest, the stability of the rate of interest must keep the level of prices fixed and stable. Stop the engine, you stop the train.

The root of the discount fallacy to which our reformers have fallen a victim resides in the orthodox notion that an increase of the supply of money reduces the rate of interest. This fundamental error permeates not only the theories of currency, but every department of political economy. I find it vitiating the standard work which more than any other has moulded, not only in England, the science of economics, Mill's *Principles of Political Economy.* I refer the reader specially to Book III, chap. xxii, § 2. But it may not be out of place to discuss the matter briefly. Mill writes :

> Suppose that England possessed a currency wholly metallic, of twenty millions sterling, and that suddenly twenty millions of bank-notes were sent into circulation. If these were issued by bankers, they would be employed in loans, or in the purchase of securities, and would therefore create a sudden fall in the rate of interest, which would probably send a great part of the twenty millions of gold out of the country as capital, to seek a higher rate of interest elsewhere, before there had been time for any action on prices.

Why should this additional money cause the rate of interest to fall ? The argument is to this effect : the rate of interest is the price of loan money ; price is determined by supply and demand ; the supply being increased, the price must fall. Mill supposes that gold would emigrate and prices remain unaffected. But he has another alternative :

> But we will suppose that the notes are not issued by bankers, or moneylenders of any kind, but by manufacturers, in the payment of wages and purchase of materials, or by the Government in its ordinary expenses, so that the whole amount would be rapidly carried into the markets for commodities. The following would be the natural order of consequences. All prices would rise greatly. . . .

In the latter case the rate of interest would not be affected at all. Mill specially stresses this point in § 4 :

> From the preceding considerations it would be seen, even if it were not otherwise evident, how great an error it is to imagine that the rate of interest bears any necessary relation to the quantity or value of the money in circulation. An increase of the currency has in itself no effect, and is incapable of having any effect, on the rate of interest. A paper currency issued by Government in the payment of its ordinary expenses, in however great excess it may be issued, affects the rate of interest in no manner whatever.

The two opposite cases are usually combined :

> It is perfectly true that in England, and in most other commercial countries, an addition to the currency almost always *seems* to have the effect of lowering the rate of interest ; because it is almost always accompanied by something which really has that tendency. The currency in common use, being a currency provided by bankers, is all issued in the way of loans. . . . The same operation, therefore, which adds to the currency, also adds to the loans, or to the capital seeking investment on loan ; properly, indeed, the currency is only increased in order that the loans may be increased. Now, though as currency these issues have not an effect on interest, as loans they have.

And as loans they depress the rate of interest, though as currency they at the same time raise prices : the rate of interest goes down while the level of prices goes up. This is diametrically opposed to what has just been affirmed. Of course Mill's theory prevails to this day, and after having set it forth we understand why Mr. Keynes seems so surprised at and so distrustful of his discovery, which contradicts it.

The theory is wrong because, in considering the effects of increasing supply, it overlooks the other side of the equation : demand. Demand does not remain passive when the supply of money is increased. It starts up furiously, ravenously, at the slightest hint of an impending increase of the money supply. For money *is* demand, demand incarnate. To the increase of money is added, doubling and trebling it, the increase of credit—credit as created by the borrowers. Everybody believes in an imminent rise of prices ; everybody is anxious to buy before it takes place ; everybody reaches out for the means to buy with, money, and instead of falling, as by the orthodox theory, the rate of interest leaps up. (More of this below.)

The quotations from Mill furnish the best clue to the misconception regarding the effects of discount policy. Since it is admitted that an additional supply of money reduces the rate of interest, the conclusions from the premise are as follows :

(1) A reduction of the rate of interest must be connected with an increase of the supply of money ;

(2) When an increase of the supply of money is wanted the rate of discount must be reduced, and

(3) Conversely, when prices have to be checked, the means to the end is to raise the bank-rate.

But in reality it is all the other way about for the reasons already indicated, which will appear all the more weighty, if we follow the error into one of its innumerable ramifications. I again quote from Mill :

> Effects of another kind, however, will have been produced. Twenty millions which formerly existed in the unproductive form of metallic money have been converted into what is, or is capable of becoming, productive capital. This gain is at first made by England at the expense of other countries, who have taken the superfluity of this costly and unproductive article off her hands, giving for it an equivalent value in other commodities. By degrees the loss is made up to those countries. . . .
>
> The value saved to the community by thus dispensing with metallic money is a clear gain to those who provide the substitute. They have the use of twenty millions of circulating medium which have cost them only the expense of an engraver's plate. If they employ this accession to their fortunes as productive capital, the produce of the country is increased and the community benefited, as much as by any other capital of equal amount. . . .
>
> The substitution, therefore, of paper for the precious metals should always be carried as far as is consistent with safety.

We are asked to believe, then, that the increase of mere money brings in its wake an equal increase of production, i.e. of real wealth. This would, to be sure, account for a fall of the rate of interest. However and again, the primary and more powerful effect is overlooked, to wit the increase of consumption and destruction. There is no such powerful destroyer as a general rise of prices, and if it does stimulate output, this output is in itself a form of destruction. When prices expand the whole body economic expands, and the effects are the same as when a boy suddenly starts into one of his growing fits: within a few months all his clothes have become too tight and too short, and have to be discarded. When business expands, real capital is discarded faster than it can be replaced; it is an orgy of destruction. If one class of the people are enriched, there is certainly an excess of impoverishment on the other side. Nothing could be more erroneous than the opinion expressed in the last quotation and the enmity to gold which it implies. There is not only no clear gain, but the losses overbalance the gains. The idea that money which is created at no expense can enrich a nation is proved to be a deception, which I have already had occasion to touch upon (p. 90). In the case of Mill it is a horrid relapse into the crudities of the Mercantile Theory, which he has so vigorously denounced.

When the doctors disagree we know for sure that they cannot all be right; but when they do agree we may be quite sure that they are all wrong. The money reformers agree in affirming that the traditional and orthodox discount practice is right: doctors and practitioners agree. By my reading of the facts that practice is the cardinal vice of the existing system. If in the present-day controversies on the subject of currency regulation there is no real advance over the writings of eighty years ago, it is because both the critics and the upholders of the system agree in attributing to a rising rate of discount a depressive, to a falling rate an uplifting effect on prices. A characteristic expression of a nascent feeling that something might be wrong about the principle I find in the utterances of those who deny to bank-rate any decisive effect whatever; for an example see the "Protokoll der eidgenössischen Währungskommission" (Minutes of the Swiss Currency Commission), Verlag des Schweizer Freiland Freigeld Bundes, Bern, p. 34–5.

IV

FURTHER UNRAVELLING OF THE DISCOUNT TANGLE

THE main thesis of this book is to the effect that the rate of interest determines the purchasing power, or value, of money. Let me open this chapter with a sentence from Mill's *Principles* :

> Almost every speculation respecting the economical interests of a society thus constituted implies some theory of Value : the smallest error on that subject infects with corresponding error all our conclusions ; and anything vague or misty in our conception of it creates confusion and uncertainty in everything else.

When, after a good deal of experimenting with my disconcerting fancies, and speculating on the possible bearings of the new principles, I had overcome my doubts as to the soundness of my own conclusions, I began to try and obtain a hearing for my case. I sent an article to the editor of an English weekly, in which the stabilization proposals of Mr. Lloyd and Mr. Keynes had been widely discussed and warmly supported. The editor returned my contribution with the remark that he could not dare to publish it because it would unsettle the opinions on the matter—formed and fostered at such great pains—in the heads of his readers : my views were altogether too heretical, and he himself was not able to credit them. And this very capable and influential friend of currency stabilization declined to be further enlightened on the subject—though I do think he has become somewhat shaken in his assurance. Anyhow, I met with experiences of a similar kind among my nearer acquaintances and former collaborators; some of them gave me up, either as a crank or as a traitor of the cause. That I was taking too much for granted in trying to win the assent of people I began to realize most acutely when I approached the truly initiated, those who are considered as the knowing ones, the experts. Well, the opposition which I encountered was in so far a help as it forced me to turn the matter over and over again. In so doing I not only was illuminated by a few new lights, but I also hit upon further methods for demonstrating the facts as I saw them. (Most of all I was gloriously confirmed in my conviction that my theory is right.) I cannot afford to withhold these new proofs from my present readers.

I

The price of goods is influenced, or rather determined, by the rate of discount in the same way as the price of real capital is by the market rate of interest. The discount rate is the rate of capitalization of the price of goods, as the rate of interest is the rate of capitalization of real capital. Capital proceeds divided by the rate of interest and multiplied by 100 = capital sum. Example: net proceeds of an estate = £40; rate of interest 4 per cent.; price of the estate = £1,000. If the rate of interest is raised to 5 per cent. without an equal increase of the proceeds, the price of the estate is reduced to £800. For the price to be maintained the proceeds must be augmented to £50. Neither will this suffice to re-establish the original ratio; the price of the estate itself must be raised in the same proportion, because otherwise the capital is depreciated in so far as £1,000, obtained when the current rate of interest is five, does not represent the same value as the same sum of £1,000 did previously, when the rate of interest was only 4 per cent. We are to understand that *a sum of*

£1,000 *at a higher rate of interest is worth less than a sum of a thousand at a lower rate.* We are up against the law of relativity.

To help the reader along with the seeming paradox I would beg him to figure the matter out in this fashion. Suppose you wish to buy a house. You are able to pay £40 interest a year, but you cannot afford to pay £50. The price of a house is £1,000. Your means, then, are sufficient to acquire the house—your money has sufficient purchasing power—if you can obtain a loan at 4 per cent., but it is without your reach if you have to pay 5 per cent. interest. At the higher rate your money—as a purchaser—is depreciated ; *the purchasing power of money is lessened by the raising of the rate of interest.*

Now let us apply this method to the trade in goods. Suppose Mercator has to pay £4 interest on the sum to be invested in the purchase of a parcel of goods. The rate of interest is 4 per cent. ; hence the price of the goods which he can buy will be £400, this being the sum which he can raise by having a three months' bill discounted. But if the rate of discount were at 5 per cent., the price of the goods would be reduced to £320, unless he contrives to increase the interest proceeds on his investment (from £4 to £5). He can charge this increase to the selling price of the goods ; but that will not suffice to make things straight for him. The price of the goods must go up in the same proportion as the interest rate, not only by the increase of the interest sum—from £400 to £500 ; he must calculate his selling price as if the buying price had been £500. This bigger sum is substantially, i.e. in its proportion to the interest paid and to its purchasing power, no more than the smaller sum of £400 was at 4 per cent. ; for the equation is like this : 5 : 500 = 4 : 400.

We are not yet fully satisfied. Let us look at the matter from the seller's point of view. Suppose I bought a house at £1,000 a year ago and now wish to sell it. If meanwhile the rate of interest has gone up from 4 per cent. to 5 per cent. I must be able to sell at £1,250. This may seem to contradict what was shown above. Who would pay £1,250 with interest at 5 per cent., when it was found that at 5 per cent. £1,000 is too big a price ? He will pay the price whose means have risen in the same proportion. This we shall understand after having answered the question why the rate of interest has gone up from 4 to 5. Why does it ever rise ? It must be in consequence of livelier demand (which may be accounted for in whatever way you please), which forces up the price of goods and real capital. This forcing up of the prices of commodities is the same as forcing down the value of money ; by the time when the price movement has reached the dimension equivalent to a ratio of 4 : 5, the purchasing power of £1,000 at the lower rate is represented by £1,250 at the higher rate. The bigger sum of £1,250 is not in substance any more than was the slighter sum of £1,000 previously. There has not been any increment of value, because £1,250 paying interest at 5 per cent. are only just as much as £1,000 at 4 per cent. The larger number at best compensates the reduced unit ; there is a greater number of pounds, but the pounds are proportionately lighter.

We have, then, passed from an initial ratio, reading like this :

The rate of interest, 4 per cent. ;
Interest proceeds (or interest paid), £40 ;
The price (capital sum), £1,000,

through successive advances of

(*a*) The rate of interest from 4 to 5 per cent. ;
(*b*) The interest proceeds (interest paid) from £40 to £50 ;
(*c*) The capital sum from £1,000 to £1,250.

What is the rate of interest, when the capital sum is £1,250 and the interest £50 ? It is no longer 5 per cent., but only 4 per cent. What has happened ? The raising of the price has brought the rate of interest down from its raised stand and re-established the original proportion. By the time when, in consequence of the rising of the rate of interest, the price has been raised proportionately, *the rate of interest naturally relapses to its initial level.*

We have here come upon a most important fact, which I believe has not yet been properly understood or taken into account. The rate of interest, that we know, has a tendency to return to a certain level. It cannot be maintained for a length of time at a point appreciably above, or below, that level, unless the level of prices is allowed to continue to rise (or to fall). It swings back to its normal level as soon as prices begin to steady themselves. Usually the level of prices goes down somewhat before stabilization takes place ; but that is a point which we need not consider in this connection.

In our hypothetical case we observe, then, these changes and adaptations :

(1) In the initial stage the rate of interest was 4 per cent., the interest paid £40, the capital sum £1,000.

(2) In the intermediate stage, the rate of interest having gone up to 5 per cent., the interest paid rose to £50, the capital sum to £1,250.

(3) In the final stage the capital sum remained at £1,250, the interest paid at £50, while the rate of interest fell to 4 per cent.

The sums having been increased in the proportion of 4 to 5, were left at the height once attained ; but *the rate of interest, after having achieved the work of inflation, relapsed.* Those £1,250 of the capital sum, and those £50 interest, as far as their substance, that is their purchasing power or real value, is concerned, are exactly as much as £1,000 capital sum and £40 interest were in the initial stage.

Before I proceed I would beg the reader to consider the actual value of his country's money in comparison to what it was in 1914. In the countries with a currency near the gold parity you have to give about £160 for what you could buy at £100 before the course of inflation. But the rate of discount in all these countries is either at the level of 1914 or below it. I shall supplement this rather general statement with a brief survey of certain special occurrences observed in the truly experimental years that have elapsed.

During the period of inflation in Germany, more particularly in its earlier stages, many inexperienced or over-scrupulous tradesmen came to grief because they sold at prices which they calculated from their buying prices, instead of by the prices which they would have to concede in re-stocking their stores. Our demonstration of the course taken by inflation showed that the rise of prices must inevitably follow upon a rise of the rate of interest: the parcel of goods went from £400 to £500 when the discount rate was raised from 4 per cent. to 5 per cent. If Mercator, having purchased it at £400, sold it at the customary money profit after three months, he would not realize enough money to buy another lot of equal value. The money which he realizes would be depreciated money. To avoid a loss he would have to raise his selling price in the proportion that the discount rate was raised. Of course if he does so raise his price he makes an extra profit for himself, because what he returns to his lender is no more than the nominal sum borrowed; and he returns depreciated money while retaining the margin. It appears from this that some one or other is bound to be cheated. The only way to avoid such unfair practices is to avoid changes in the discount rate.

It is quite obvious, then, that *the rate of interest constitutes the standard of money value.* Variations of the rate must needs alter the purchasing power of money and cause the evils inherent in such fluctuations. The rate of interest governs all the economic relations. A sum borrowed at 5 per cent. is not the same value as an equal sum borrowed at 4 per cent. Suppose I have to pay £100. I can raise the sum either by selling some property, or by obtaining a loan. In the latter case I purchase the use of the money with the annual interest which I engage to pay. The lender purchases with his money this annual tribute, while I purchase the use of the money with the annual tribute, called interest. The interest which the money-lender receives is probably spent for the purchase of commodities, and so we may say that interest is commodities at one remove. *Thus interest and commodities are seen to stand in exactly the same relation to money*; the value of money is low when commodity prices and interest are high, and when interest (the rate of interest) and commodity prices are low the value of money is high. The case of money and commodity prices has come to be fairly well understood by most people, whereas the case of interest is still looked at upside-down.

I do not believe in the superior merits of demonstration by statistics. As a general rule figures rather blur the survey. In presenting the case of the Swiss currency during the period since the outbreak of the war I shall beware of being too accurate or going into unnecessary details. I choose the Swiss record not only because the data come to hand more readily, but chiefly because no other country shows so clear an outline.

My theory contends that interest determines prices and must therefore precede price. The practical experiment would, therefore, begin by a move of the rate of interest and then let things take their natural course. The Swiss experiment went off as follows:

The discount rate of the Swiss National Bank was 3½ per cent. from February 19 to July 28, 1914; it was moved to

4½ per cent. on July 28th
5 per cent. on July 29th
5½ per cent. on July 30th
6 per cent. on August 2nd
5 per cent. on September 9th
4½ per cent. on December 30th

Obviously the experts who decided upon these hops and skips must have known what they were about. We observe, then, a sudden and powerful raising of the discount rate. (To cumulate the effect a Federal Loan was issued at 5 per cent., a rate far beyond the rates prevailing at the time, within the first fortnight after the beginning of the war.) What may have been the reasons of such policy ? As a protection of the gold reserve it cannot have been intended, since the obligation to redeem notes was suspended. The only possible purpose must have been to raise a barrier against an upward movement of prices—for, surely, nobody could have thought of throwing a special obstacle in the way of enterprise. If a check on prices was the object pursued, the means employed must be pronounced to have miscarried : the effect produced was all in the opposite direction ; it was not long before prices began to move upwards.

The raising of the discount rate had created the scope for, and given the initial impetus to, the issue of a very considerable quantity of new bank-notes and currency notes. Its immediate effect, according to my theory, was to diminish the quantity of the money substance outstanding, because money loses substance when the rate of interest goes up. The pinch was immediately felt by the business world ; people who depended on being well supplied with money were forced to obtain larger sums. Every one tried to get hold of a little more than he needed at the moment, and those who were possessed of sums were careful not to let them escape. All these are the inevitable reactions upon a diminution of the money supply. The consequence was that money had to be newly created.[1] By December 1st

[1] The main agent of inflation was, however, the expansion of credit which took place, and was bound to take place, under the circumstances. When the discount rate is raised above normal the business world always have recourse to self-created credit instruments, which become safe when prices go up. And prices invariably do go up when credit transactions increase. Prices go up : this signifies that money is depreciated. Such

the note issue had increased from 275 millions to 414 millions, i.e. some 60 per cent. (Very characteristically the publications of the Swiss money reformers point to no connection between the movements of the discount rate and the note issue; they do not so much as indicate the former. Had they heeded the parallel they could not have helped noticing how contradictory their discount theory was.)

After the train was thus started it moved on and on. The discount rate was left unchanged all through the following four years; not till October 18, 1918, was it moved again. The war was decided; new developments must be looked forward to. What could the expectations be in the matter of price? Better supplies, an easing of the situation: such were the general hopes. The index of the cost of living had risen to 248 from its base of 100 in July 1914. It was necessary to check the race to the high mountain-tops, and, agreeably to the teachings of the accepted theory, to achieve the end the discount rate was raised from $4\frac{1}{2}$ per cent. to $5\frac{1}{2}$ per cent. Here is a passage from the book of one of the directors of the Swiss National Bank explaining the experiment:

> The discount policy of the central Bank of Issue will then take measures; by raising the rate of discount it will hamper the supply of cash; the bank will speed up the withdrawal of notes. The reduction of the currency in circulation will reduce purchasing power,[1] keep the demand for goods low, and so depress the price of goods.—*Krieg und Geld,* by Ott. Bern, 1916.

The discount rate was raised with a view to bridling the career of prices. Alas! things happened agreeably to my

depreciation could not happen if bank-notes were as eagerly sought after (appreciated) as the text seems to imply. By the free use of bills of exchange trade is enabled to dispense with, to do without and neglect, bank-notes, and it is this neglect of cash money which is the real cause of its depreciation. At the same time this depreciation is tantamount to a diminution of the total quantity of money, and naturally induces new issues of bank-notes.

[1] The term "purchasing power" is here used in the sense of volume of bank-notes issued.

theory, not the director's. The Index of prices grew all the more vigorously along with the issue of bank-notes. It kept advancing, with only one brief interruption, up to the second half of 1919, when a great general strike occurred as an expression of the people's wrath.

Again the experiment might have proved to the experts, theoretical and practical, that the accepted discount notions must be wrong. It would seem as if the Swiss Bank had heeded the lesson. On August 21, 1919, the discount rate was reduced to 5 per cent. That was the strangest, the maddest move—considering what the official theory was—that could have been made. It was unique, not imitated from any other country, not imitated by any other country. And its effects were truly remarkable. The Swiss price movement detached itself from that of all other currencies : while in all other countries the price Indexes continued to rise, the Swiss Index henceforward moved in a horizontal zigzag, rising and falling slightly, and alternately from month to month. No other bank of issue, so far as I have been able to ascertain, had reduced its discount rate. The Swiss National Bank did not institute any new methods, either. So the only difference consisted in this lowering of the discount rate. To what else could the independence of the Swiss Index be attributed ?

The great world crisis began to make itself felt in Switzerland towards the end of 1920. Influences were at work that overrode the discount factor (see above, p. 125). Very soon, however, the discount began to play again. It was lowered to 4½ per cent. on April 7, 1921, and after that consecutively by one half per cent. down to 3 per cent., which lowest point was reached on August 17, 1922. I purposely avoid adding the Index figures. Along with the discount they declined fairly steadily, until they touched the bottom line about the same time as the discount did.

The Swiss experiment demonstrates and proves, supposing that observation and measurements and statistics do prove anything, that (*a*) The price level rose when and while the discount rate was raised and kept at a high figure. The variations of the discount preceded, but the variation of the index figure went to bigger dimensions. (*b*) The price level fell when and while the discount was reduced. Contrary to (*a*) the reduction of the Index did not proceed as far as that of the discount rate.

My theory, then, is perfectly confirmed. A rising rate of discount forces up prices (reducing the rate of exchange and finally driving gold over the frontier—provided that there are countries which are less afflicted); a falling discount rate depresses the price level (raising the rate of exchange and finally attracting gold to the country in question).

I wish to qualify these statements. It is not really the raising and reducing of the discount rate that produces these results; life is not so mechanical as that. But the movements of the discount have so far been the expression of natural tendencies, and they have been so handled as to stimulate these tendencies—which was the great mistake.

II

The nature of interest will be touched upon in a later section. From what has been said in this chapter it clearly appears that the rate of interest is an expression, or a translation, of the people's interest in real values. When many people, at a given point of time, manifest a strong interest in and create a heavy demand for goods and real capital, the first outward sign of it appears in the form of rising rates (borrowing rates rather than lending rates) for *money*, because the demand for goods can only become effective through money. The initial move and the real reason is an interest in commodities, which, however, cannot translate itself into effective demand so long as the money is not at hand. First of all people have to procure money (either cash or at three months' date through credit); to obtain it they are willing to concede a higher rate. The purchase of real values follows suit, but it is obvious that the heavier demand for money, and the consequent rise of the rate of interest, precedes, and brings in its wake, the heavier effective demand for real values and the consequent rise of prices. If we agree to designate by the term of inflation this general enhancement of price, we shall also agree in attributing to the raising of the rate of interest the cause of such inflation. When the level of prices goes up it is because a strong, an excessive interest is at work; this interest must be considered as the primary factor in the chain and network of influences which produce the final result. The rate of interest is the arithmetical manifestation of this interest.

There is a notable difference between the movement of the rate of interest and the level of prices. Whereas the rate of interest moves by sudden leaps, the level of prices shifts by gentler degrees, though by no means in a steady ascent or descent. Half a per cent., an advance from 4 per cent. to $4\frac{1}{2}$ per cent., from $4\frac{1}{2}$ per cent. to 5 per cent., makes a difference of one-eighth, one-ninth, one-tenth of the whole, i.e. $12\frac{1}{2}$ per cent., 11 per cent., 10 per cent. Commodity prices cannot move so suddenly, much less can the general level of prices, in which all sorts of compensating movements will retard the development. Hence a period of inflation is invariably ushered in by an advance of the rate of interest, seconded by a raise of the discount rate. The subsequent raising of the level of prices is spread over a certain lapse of time. The movement may be carried on into a second, a third, a fourth stage, provided that the forces stimulating the interest of consumers remain active enough to cause the rate of interest to stay high or take a new leap after the first has been compensated by the advance of the price level. However, the experiments that we have seen carried out present a curious uniformity in showing that the rise of the Index-numbers goes by fits and starts. The actual raising of the discount rate, as a rule, is immediately followed by a slight drop of the Index-number ; this, of course, is fully accounted for by the money shortage caused through the depreciation of money in consequence of the raising of the discount. After a while the turn sets in, prices take heavy strides upwards, until the impetus grows weaker again and the advance of prices finally comes to a standstill.

What must, logically, happen when this point is attained ? We found that at a given moment the rate of interest was back again at its initial level, while the price level might be maintained at its elevation. How is this to be accounted for ? The rate of interest, it has been admitted, went up because the consumers manifested a strong interest for real values, and made a heavy effective demand for money capital in order to satisfy their interest. When the tide sets in the opposite direction it must be because the interest and the desire to buy has been satisfied or quenched, or because the capacity to buy has exhausted itself. What more natural than that the rate of interest should translate this altered state of mind into a reduction ?

But here we are faced with a complication. As a rule the first signs of this reversal are not falling rates of interest —they are rather apt to go up another bit—but the fall of prices. The rate of interest is kept high, though not from any excessive desire on the part of business men to acquire real values—desire is all the other way—but from the urgent need of the owners of goods to preserve their balance. However, before long the rate of interest, too, gives way, and if it lagged a little at first, its fall in the long run becomes all the greater, so that it is certainly correct to say that the fall of prices is induced by the fall of interest, in exactly the same manner as the rise of prices was induced by the rise of the rate of interest.

Our new conception of the effects produced by the movements of the rates of interest and discount enable us to account for certain phenomena that have never been satisfactorily explained. If it is correct that the raising of the rate of interest depreciates money, it follows that it diminishes the quantity of money outstanding. The quantity, I say, not the number of units. Every unit being diminished, the aggregate quantity must be reduced in the same proportion. This furnishes a clue to the curious belief, so universally held, that a rise of the rate of interest hampers business by making for a restriction of credit. Something does happen, as when an optical illusion produces itself; the illusion consists in the cause assigned. The measure is indeed felt as a pinch, which is due to the fact that it does diminish the available quantity of money—through the shrinkage of every money unit, as explained above. Business feels cramped for the time being, but it also feels prodded. It reacts, as it always does when an article is supplied in insufficient quantities: it makes a stronger bid for as much as is to be had; it forces the production of it. In our case: the tightening of the discount valve is invariably followed by a feverish demand for money and by a more copious issue of notes by the Central Bank. This has already been hinted in the survey of the Swiss record. I will add two more observations to the point.

During the period of inflation, both in Germany and Austria, there were occasions when money suddenly ran short or gave out, so that it was quite impossible to meet demands. It will be found that these emergencies took place whenever a higher rate of discount was declared. Quite naturally so. By the raising of the discount the money in circulation was suddenly diminished; it became insufficient to do the work required. Those who had got hold of the cash retained it, anxious to keep themselves supplied against the day when they should need it. Those who needed money grabbed more than was really required. The consequence was that the shortage declared itself in menacing proportions, and that the printing press was made to spin double quick. Thus it appears that the printing of notes was not the original cause of the money depreciation ; it came in consequence of, and as the natural (though ineffectual) remedy against, the disturbance produced by the raising of the discount rate.

During the period of the late economic crisis enormous quantities of idle money accumulated at the banks, though hardly any or no interest was paid on deposits. This is a phenomenon observed in every crisis. Silvio Gesell, championing the theory of the bi-metallist school, which had fought the crisis following upon the demonetization of silver in the seventies and eighties of last century, insists that the crisis is always due to a shortage of money. The money lies idle, he argues, because there is too little of it. The crisis alluded to was the result of the diminution of the output of gold—plenty of statistics to prove the fact—and the simultaneous demonetization of silver : there was tangibly less money available ; co-ordinate the two observed facts and the conclusion imposes itself. Now find the logical reason. It may be argued that money, being round, behaves much like the wheels of a cart ; when one wheel comes off the other three become useless : wheels have to lie idle because there is one too few. Or : if 100 hands are needed to run a factory, work must cease and the workers go idle if there are only 70 of them. However, that is not the true aspect of the case. The money lies idle because nobody has any use for it, seeing that nobody wishes to acquire real values. Money must be looked upon as a commodity. Now, it has surely never happened that demand could not be got on foot for an article that was felt to be short. On the contrary, demand grows urgent and reckless so soon as a shortage begins to loom up. I have already referred to the efforts made by the German business world in those emergencies of the *Geldknappheit*—the dreaded money shortage.

Let us, then, admit that the piles of money waiting for employment during the slump were the sign of a real superabundance of cash, caused by the swelling of the money substance owing to the dwindling of interest and the reductions of the rate of discount. This leads to a more plausible explanation of the attitude of business. The article which appears in increasing quantities does not attract the buyer, it scares him away. Why borrow money to-day if by deferring the loan a few months you can borrow better money, money that buys more ? Whether there is too much or too little money can be ascertained only in one way : by comparing the supply with the demand. As demand is lessened through the decline of interest (demand and interest are equivalents), it follows that reductions of the lending rate are bound to swell the money supply ; for, no matter what the figures of the case may be, lessening demand is the same thing as growing supply. As I said before : we are up against the law of relativity.

The real money shortage, then, is caused by inflation through the blowing up of the discount rate. But it is invariably counteracted by fresh issues of money, so that after a spell of fluctuating, or even falling, prices the Index will leap up again. In a sense it may be said that prices diminish as they rise. The higher price is the lighter price, if it comes in consequence of inflation through interest. Every fresh advance brought the German competitor ahead of his Swiss or English rival, while he might be overtaken and outstripped when the mark kept steady for a while or gained a few points. How could this have been if the higher price had also been the bigger price ? This observation may count as a further proof of my contention that a rising discount, or interest, rate diminishes the money supply and along with it price. The quantity theory is not invalidated by my reversing the relations. In like manner the swelling of the supply of money through the fall of interest rates produces what the quantity theory demands : the increase of prices. The figures shrink, to be sure, but the smaller sum covers a more massive and a heavier substance. It becomes harder for almost everybody to pay the lower price.

I have uttered a caution against the belief that the movements of the rate of discount are the primary factor. On the other hand, it may be argued that the general price movements are largely induced by forces that can be checked or turned another way by proper management of the outward agents. Beliefs, expectations, speculations, fears, hopes, are the real makers of price : interest is the term which covers them all. To canalize interest, to preserve it from extravagances, must be the aim of a currency policy aiming at stability. One of the forms of interest is the rate of discount ; therefore the control of this force must include the control over prices. Theoretically it must be possible to master the price movement through the rate of discount.

III

A version of the foregoing argument was submitted to an economist of international reputation. This great authority declined to meet the author for a discussion of the matter, on the plea that the propositions were too much beside the mark. However, the professor condescended to administer a bit of lecturing, so that I am given the happy chance of anticipating the criticisms which this book will elicit from the learned colleagues of my lofty mentor. Says Sir Oracle :

" A high rate of interest has always been considered by 'practice' the way to strap down and limit demand and enterprise, while a low rate has been looked upon as stimulating demand and enterprise. . . . You will not discover the theorist willing to concede the possibility of a discount policy altogether independent from, nay, diametrically and deliberately opposed to, the movements of the generally ruling market rate of interest."

"You propose to reduce the rate of discount when the market rate of interest tends upwards, and to raise the discount when the market rate goes slack. Practice will respond to your proposal in the following manner: If the Bank's lending rate is eased at a time when the general market rate of interest is rising, the inevitable consequence will be that within a short period the Bank's stock of notes will be exhausted, and, with no metallic restrictions on the note issue to bring the movement to a standstill—as by your scheme—the end will be a damaging course of inflation. And vice versa: if the Bank were to raise its lending rate at a time when the market rate declines, the natural reaction would consist in leaving the Bank high and dry, abandoned by business."

My humble answer to these reminders was to this effect:

All economic theory maintains, and all economic practice pretends to admit, the truth of the contention, that a high lending rate for discounts hampers business. Surely this is to affirm more than anybody will be able to prove. When does the lending rate begin to be "high"? Look at the rates actually (January 1925) obtaining. They are 4 per cent. or thereabout in the countries whose currency stands near a parity with the price of gold. But it is 7 per cent. in France, 10 per cent. in Germany, 13 per cent. in Austria. I suppose you will consent to call these rates high. Well, what do we observe? Have they been a check on business? The annual report of the Banque de France for 1924 makes this confession: "The monthly maxima of the note issue during the second semester evince a steady tendency to increase, and this is a circumstance which must *donner à penser*, cry out for a remedy. The measures so far applied by the Bank to counteract this tendency have evidently not been sufficient. The rate of discount was raised to 7 per cent., which, at the time, was strongly deprecated as likely to paralyse business ; but this is not an upper limit never to be surpassed." (Quoted from the *Basler Nachrichten.*) In the Germany of the paper money period the rate of discount rose to 1,000 per cent. ; but the greater the leaps it took the more greedily and furiously did demand manifest its needs. Even in the Germany of the so-called stabilized mark, at a time when every effort was made to restrict the issue of bank-notes, and the claims of interest were truly exorbitant, demand broke down this barrier, the note issue increasing heavily and prices advancing in a marked degree. The same observation was made in Austria : the note issue expanded when the rate of discount was 12 per cent. ; it did not cease to expand when the rate was raised to 15 per cent. ; and prices kept moving upwards all through the year of currency "stabilization" through scientific discount policy. What is to be said of a theory which, in the face of such object lessons, persists in attributing a trammelling effect to a "high" discount rate ? What are we to think of a practice which, in spite of such experiences and failures, persists in the attempt to restrict the circulation of money by means of heightened money rates ? The innocence of the Banque de France's report is not to be outdone : "evidently not sufficient." Cut off two bits and still too short ! [1]

There is your theory, there is your official practice. But

[1] When I wrote the above paragraph I knew that inflation would merrily continue in France ; but I could not have dreamt in what a spectacular fashion my expectation would be fulfilled. Within two months the Cabinet of M. Herriot was swept from office because it was forced to confess that it had induced the Banque de France to issue notes beyond the legal limit and was contemplating the issue of four more milliards. Of course the Minister of Finance made out a beautiful case of how the level of prices would not be affected by the measure, because and because and because : the arguments might be suitably inserted in *Alice in Wonderland.* The untoward development was the effect of that ill-advised, although "scientifically" correct, raising of the discount rate. M. Herriot has been hit hard, and he has cried out over machinations of high finance, this most convenient of bogies.

Here is confirmation of my interpretation from the opposite side. On everyday, unofficial, practical practice goes by better rules and laws. It sticks to a well-borne-out apophthegm of business experience, saying : *La hausse amène la hause*—the boom brings on the boom, and nothing succeeds like success. And it is on this homely wisdom of experience that my solution of the discount problem is founded.

La hausse amène la hausse : it goes from step to step. First you raise one foot—the discount foot—and then you draw the other foot—the price foot—after it, whereupon the first will again move on. You scale the heights : can you visualize it, sir ? Now, my conclusion was—and I found it ever so close at hand—that if you do not want to go upstairs, you must refrain from drawing the second foot up after the first has been tempted, or betrayed, or forced, on to the first step—still less should you raise the discount rate deliberately, in order to forestall the rise of prices, as was done in 1914. The rate of discount is a price, and so, if it is true that the boom brings on the boom, the way to inhibit the boom, to nip him in the bud, would seem to consist in taking up a firm stand at some decisive point : refuse to budge. You, sir, believe that it will not be feasible to make the Bank's discount rate detach itself from the prevalent trend. I shall answer your reasons more fully below ; but it is a *prima facie* conclusion that,

November 6, 1924, the Austrian National Bank reduced its rate from 15 per cent. to 13 per cent. By the end of April 1925 the amount of discounted bills was halved, the note issue reduced by one-tenth, and the gold reserve increased by one-tenth. When the reduction of the rate was declared, the London *Times* thundered at the unhappy Austrians, charging them with making deliberately for more inflation. However, it has resulted in deflation. Business in Austria all the while has been going extremely slack. But logic and sound science are coming to its rescue. It was the reduction of the discount rate which caused the general contraction ; well, the Bank now proceeds to a further reduction of 2 per cent. Its benevolent intention is to procure business "cheaper" money ; but it will only succeed in strengthening the prevailing tendency. Of course the Austrian rate will eventually have to come down to a more normal level. However, leaps of 2 per cent. at a time are too shocking, and it will require a better science before the transition can be successfully managed.

from the moment when the Bank of Issue has engaged to conduct business with a view to stabilizing the currency, it cannot logically follow suit with the fluctuations of the market. *Its action must be a counter-action, which means either action in the opposite direction, or non-action.* These are conclusions which a man requires no expert knowledge to arrive at. Take a long breath, use your natural logic, sir, and you will be rid of your mildewed theories. And has it not struck you how the rates of interest have fluctuated these last eighteen months, here in our Switzerland, while the rate of discount was kept unchanged ?

La hausse amène la hausse, success succeeds, says the adage. The practice of the Banks of Issue so far has proceeded—and the theories of the learned experts have argued—as if the boom brought on the slump, as if success were bound to fail. Of course this statement is also true. When the boom has run its course and had its day, the slump will make its appearance, and as the boom has preceded, it is correct to say that the boom brings on the slump. Now the question is : which of the two conflicting interpretations must be adopted as a guide for our measures to stabilize the currency. This question we shall decide by discovering which of the two can produce practical results.

One sentence says : *la hausse amène la hausse*, and says too much. For in the long run the boom comes to an end, and the last raising of the rate of discount will be followed by the decline of prices. Only you never can tell which is going to be the last—I know one who was named Benjamin, because he was to be the last, and who has no less than three younger brothers. Therefore it also is correct to say that the *hausse* brings on the *baisse*. However, this, too, is saying too much, since it is only the last *hausse* which ushers in the *baisse*. A very great teacher of science, Ernst Mach, tells us how natural laws are established. " By their origin natural laws are limits which, guided by experience, we prescribe to, or impose on, our expectations." In the present instance the limit imposed on our expectations will reduce our statement to something like this "natural law":

(*a*) *In its beginning* (incipient stages) the *hausse* produces a *hausse*—a gradual, consecutive raising of prices, one move always inducing the next move: from discount rate to price, from price to discount rate.

(*b*) *At its close* the *hausse* swings round and turns into a *baisse*. Applied to the theory of discount this law will teach us that the first, and second, and third (why not the ninth and the tenth?) raising of the discount rate carries the boom of prices farther and farther from level to level; but the last raise is succeeded by a reversal, though, of course, it is not this last straw which breaks the camel's back.

The case under (*b*) then marks what in German is called a *Grenzfall*; it is not good to go by, because it says no more than that the life of the most successful and healthiest person must end in death. Policy must be based on the main case. Hence a discount policy devised to forestall fluctuations of the level of prices must not encourage, and so to speak finance, the impending boom of prices by raising the rate, but rather it must counteract the movement in its incipient stage by keeping the rate low. The discount slump will bar the course of the price boom, strangle it in the cradle. For if the sentence holds good that *la hausse amène la hausse*, it must also hold good when expressed in negative terms: *la baisse ramène la hausse*—the slump brings down the boom.

It stares you in the face, learned sir. Can't you see it? Does it not startle you? You will not make it out with a telescope, nor with a microscope either; but plain eyes will do the job nicely if you open them unflinchingly. The people who manage to amass fortunes by speculation go by such everyday, commonplace rules as this one is. They may sound paradoxical or contradictory, these rules, when set off against your fine theories of the schools ; but a little bit of deliberation, a moment's fixing of the object with eyes freed from blinkers, and you cannot help grasping the better truth. Indeed, I am at a loss to comprehend how it is that I should be spending my time, so very late in the day, after so many have gone before, in trying to make you grasp a fact which cannot but come to be felt a truism, when it has once been discovered.

You have read my paper since you criticize it. I will be forbearing and try to explain once more what has already been explained, in order to allay your fears regarding the practical effects of my discount policy. You warn me that the issue of notes would grow and exceed all bounds, if the Bank grants easy lending terms at a time of rising market rates of interest. Have you been awake these eighteen months ? Our rate of discount stood at 4 per cent., the ruling market rates were never below 5 per cent., and for quite a period they were 5½ per cent. : have you noticed that the note issue showed any tendency to increase, the way it did in France, Germany, Austria ? Does it not strike you, sir, that the market rates might be prevented from rising, nay, reduced, if the official discount rate were kept low ? Suppose the Swiss Government, determined to maintain a stable price of wheat, proceeded on the lines indicated by my scheme. It provides an inexhaustible store of wheat against an emergency. When for some reason or other the market price of wheat begins to rise, the Government stores are offered at the former price : do you think, sir, that the price of wheat will keep rising ? It certainly will not. Can you imagine the effect of such action on the corn exchange ? It would mark the end of gambling of the kind that we have been witnessing these days. Bank-notes are bought and sold (demanded and supplied) in exactly the same way as wheat. When it is known that there is an unlimited supply surely nobody will think of taking out more than he has a profitable use for—considering that *there is a price to be paid*, a fixed price at that, which will not be raised to the later comers. Why should I borrow in a hurry if I am assured that the terms will be the same next week, and next month, and next year ? Here again ordinary logic suffices to get at the truth of the matter. Besides we have enjoyed lately so many fine object lessons to teach us that simple truth. Did you see people rushing out to clear cute Mr. A.'s shop when he announced a cut of prices ? Certainly not. They all strolled round the corner to find out whether competitive Mr. B. had not yet cut a bit lower. We want our Bank of Issue to be the competitor of the other money lenders. When these would run a race uphill our Bank notifies its intention to stay in the plain, which will stop the race beautifully. When one foot refuses to stray, the other must needs keep to the road.

This book contains some very severe criticism of Silvio Gesell's errors. When, from testimonies like our great man's objections, I realize at what a very low ebb the science of our faculties of economics stands, I recover some of my former admiration for the achievement of the solitary seeker and adventurous pioneer Gesell. The point which I have been stressing in this prolonged discussion of the discount problem was first made by Gesell when he demonstrated the mechanism of his artificial *Schwundgeld* twenty years ago. It is proposed to fix the rate of depreciation of this hypothetical money at 5 per cent. per annum. Now this depreciation is intended to supply the function performed by interest in a traditional currency system, and Gesell suggests that one way to influence the circulation of money so as to insure stability would be by varying the rate of depreciation: *raise* it when prices *decline*, *lower* it when they *rise*. That is exactly the same thing as raising and lowering the discount rate. I need not enter into any conjectures as to the reasons which may have prevented Gesell from detecting his discount fallacy. The truth is that he has never corrected it, nor ever objected to the way in which his chief interpreters have handled the question. In one of the newest publications to show the superior merits of the Free-Money system, you may read, on one and the same page, these two rules for steadying a failing price-level: (2) "*Reduce* the rate of discount." (6) "*Raise* the rate of depreciation, by which a speeding up of the supply of money is brought about."

V

MONEY AND CREDIT

In his very illuminating chapter on *The Regulation of a Convertible Paper Currency* Mill states the case of the relation of money to credit as follows:

> But when the value of a metallic or of any currency is spoken of, there are two points to be considered; the permanent or average value, and the fluctuations. It is to the permanent value of a metallic currency that the value of a paper currency ought to conform. But there is no obvious reason why it should be required to conform to the fluctuations too. The only object of its conforming at all is steadiness of value; and with respect to fluctuations the sole thing desirable is that they should be the smallest possible. Now the fluctuations in the value of the currency are determined, not by its quantity, whether it consist of gold or of paper, but by the expansions and contractions of credit. To discover, therefore, what currency will conform the most nearly to the permanent value of the precious metals, we must find under what currency the variations in credit are least frequent and least extreme. Now, whether this object is best attained by a metallic currency (and therefore by a paper currency exactly conforming in quantity to it) is precisely the question to be decided.

This is most excellently conceived, yet, when closely scrutinized, it is found to avoid the point at issue. If it is credit that causes the fluctuations, and if it is these fluctuations which in their turn induce the changing moods of credit, it is evidently immaterial which course is chosen, because we are in a vicious circle. Sheer common sense would suggest, it seems to me, to go at the responsible factor direct and just prevent it in the most uncompromising and brutal manner from fluctuating : strap him down. This agent is admitted to be credit. Credit manifests itself in the rate of interest ; the fluctuations of credit are manifested in the fluctuations of the rate of interest. Therefore strap down the rate in that one of its forms which is accessible to official interference : the rate of discount. That is what my system suggests.

At the bottom of the disagreements in the money reform schemes lies the problem whether money or credit is the primary factor ; quite naturally the solution proposed depends on the position adopted by the investigator. Mr. Keynes seems to have somewhat shifted his stand. In his book he expresses scepticism as to the scheme of Professor Fisher, which relies on the material or money factor : " In any case I doubt the wisdom and the practicability of a system so cut and dried." He favours the control of credit : " Thus the tendency of to-day—rightly, I think—is to watch and to control the creation of credit and to let the creation of currency follow suit." While Professor Fisher, in the meanwhile, has given in to Mr. Keynes's criticism and come to admit that along with the material factor of money it is necessary to control the credit factor, Mr. Keynes has moved over to Professor Fisher's point of view. His conception of the relation of credit to money is most succinctly stated in his *New Leader* article :

> One method is by adjusting the bank-rate. I personally have come more and more to the view that the more powerful method—certainly the weapon upon which one would have to depend in the last resort—would be variations in the volume of deposits in the Bank of England available to joint-stock banks as the basis of their working credit.

If I am not mistaken the deposits in the Bank of England represent money, cash. They are the *n* (" currency notes or other forms of cash ") and *r* (" the proportion *r* of their potential liabilities which the banks keep in cash ") which, according to the passage quoted above (p. 121), " we must be prepared to vary on occasion." So then "the more powerful method" would seem to consist in managing money, "the basis of their working credit." And this is what the article suggests as to the procedure :

> Now, there are two ways in which the amount of credit they offer can be reduced ; partly by raising the price—that is the method of the bank-rate—and partly by diminishing the supply or increasing it, as the case may be, which can be effected at one remove by varying the amount of the deposits of the bankers at the Bank of England. That is a matter which is just as much under the control of the Bank of England as the bank-rate is, because by buying or selling conversion loan or any other security they can reduce the bankers' balances or increase them within considerable limits at will. I believe the limits are quite big enough to produce a drastic effect, so that the actual technique of the business would be by expanding and curtailing the basis of credit quite as much as by raising and/or lowering the price of credit.

Mr. Keynes is the recognized authority on the problem. All the more necessary is it, therefore, to be sure that we read him rightly and bring to light any errors that we may suspect in his argument. I would rather agree with him and have his support, than criticize him and risk his opposition ; but I cannot help thinking that his view is untenable. What does his "partly . . . partly" signify but that the two ways proposed complement each other ? But do they ? Re-read the last sentence of the quotation : it implies that the price of credit can be managed independently of the basis, i.e. the volume or supply, of credit. It is a repudiation of the principle that price is determined by supply. I call that managing with a vengeance ; it will require powerful managers to achieve the end.

The vice of the argument is clearly betrayed by the opinion that the Bank of England can buy and sell securities at will. Nobody can buy or sell at will. In the first place the Bank can only sell securities if it possesses some ; it can only buy them if others possess some. In the second place the Bank depends on the willingness of another party to sell when it wants to buy, and to buy when it wishes to sell. Securities are bought and sold according to the profit they yield, i.e. the rate of interest, and so it depends on the bank-rate whether the deposits at the Bank can be increased or diminished. If the rate does not induce the public to buy securities, no securities can be sold, and the volume of money in circulation cannot be diminished to check the rise of prices; when the rate does not induce the public to sell securities, no securities can be bought, and the deposits of cash cannot be used to augment the circulation.[1] The rate of interest governs, sovereignly, the flow of cash to and from the Bank's deposits, and the "partly . . . partly" is reduced to the one and sole factor of credit. It is credit, as expressed in the rate of discount, that makes and unmakes cash. What is right in the method suggested by Mr. Keynes is that it provides (1) that the Bank shall keep at the disposal of the public as much cash as may be called for, and (2) that the Bank shall receive back its cash when the public have no use for it. Mr. Keynes wants the Bank to manage, to act, to take the initiative; but the Bank cannot act, the initiative is with the borrowers, and the Bank must be passive.

I cannot imagine what has led Mr. Keynes to conceive his double measure, which, on the face of it, runs counter to the best methods of science. He invariably relinquishes one of his positions when it comes to indicating a positive policy. Thus in the article here considered he is reduced to bank-rate in his advice how to meet the new boom which he anticipates. His distrust of the efficacy of bank-rate policy can only originate in a suspicion that the policy advocated by him is not the appropriate one.

Does credit create money, does money create credit: that is the question. Let us consider in some detail the case

[1] When prices tend to rise nobody will buy securities; when prices tend to fall everybody wants to buy securities. Thus the currency notes refuse to return to the Bank when they most ought to, and they return most eagerly when they ought not to.

which Silvio Gesell makes out for the primacy of money. He writes in *Aktive Währungspolitik* (p. 12):

> It is true that bills of exchange and cheques have partly replaced cash; but has not the economic importance of money been rather enhanced thereby? Is not the bill of exchange, the cheque, the whole fabric of credit based on cash; are not all liabilities and securities expressed in terms of money, and does not the whole structure of bills of exchange, securities, State bonds, stocks and shares collapse, when its foundation, which is cash money, is removed?

This statement leaves no doubt as to the precedence of money. However, Gesell knows of something that was before money. I will quote more: "Division of labour demands an exchange of products and depends on it. It can only proceed so far as the exchange of products will permit. The limits of division of labour are drawn by the difficulties of the exchange. . . . The main difficulties of commerce were removed and overcome by the help of money. Without money the spread of division of labour would have found a barrier in the difficulties of barter. Man began to make history only when the introduction of money enabled him to rely on division of labour. . . . Those products of division of labour which are still waiting to be exchanged, i.e. which have not yet reached their consumer, are called wares. It is these wares, which depend on money for their turn-over, that gave rise to money. The more division of labour expands, the more wares does it bring to market, and the more money is required. . . . (This is to overlook) the fact that wares are the primary factor and that money (the instrument of exchange) is the derived factor. In the beginning was division of labour, which begat wares, which, in their turn, gave birth to the need of money, of an instrument of exchange."

Note how the argument turns in a circle. In one passage (p. 12) we are told that money is the essential condition and foundation of division of labour; in another (p. 15) division of labour creates, first wares, and through wares money. Our problem, however, is to know the relation of money, not to wares and division of labour, but to credit. My opinion is that it was, and is, credit which gave rise, and gives rise, to division of labour. Men began to produce wares, i.e. things not intended for their own use and consumption, because they believed that they would be able to exchange their products for the products of other men. This belief is credit, the very essence of credit, the whole of credit. If we speak of cheques, bills of exchange, securities, paper wealth generally, as credit, we take the form for the substance. There was a stage in the evolution of monetary methods when bank-notes were considered as credit, not as money; a stage may be reached when "value-papers" will be treated as money pure and simple. This kind of credit has never behaved otherwise than money. But take the word "credit" in its primitive connotation, and it becomes clear that credit, the personal, the spiritual factor, must be looked upon as the primary one, as the creator of both wares—things made for exchange against other things—and money.

I do not believe in evolution from beginnings *ab ovo*. Credit was not evolved. It is a human element, the same now as ever. I will let Silvio Gesell demonstrate how credit is the creator of wares and money to-day. In discussing the policy of the Banks of Issue he traces the causal connection from the belief in a rise of prices to the issue of new purchasing power and the actual advance of prices—from credit to money.

The merchant (manufacturer, speculator) can employ (invest) money, also borrowed money, advantageously so long as he hopes to be able to sell what he buys with the money borrowed before the maturity of the bill, and at a profit. This sole condition of commercial demand for money is fulfilled so long as prices are rising, which in its turn is bound to take place so long as the supply of money is augmented, either through the issue of bank-notes or otherwise. With the money which the Banks of Issue create and circulate by discounting bills they increase the demand for wares, and this increasing demand will naturally raise prices.

What, then, is the origin of the movement? A hope, the belief that prices are going to rise. To make his point the better understood, Gesell proceeds on this hypothesis:

Suppose the intelligence is all of a sudden published that a new goldfield, of unheard-of promise, has been discovered; suppose this report is confirmed by the arrival of a shipload of gold. Would such an intelligence, which opens a prospect of an imminent and powerful expansion of the volume of money, diminish or multiply the demand for credit at the banks of discount? . . . It is evident that every one will count on a new boom, that every one . . . will evince a lively desire to buy, that every one, eager to make the most of the boom, will buy as many wares, stocks, shares, raw material, real estate (everything with the sole exception of securities with a fixed rate of interest) as his own ready money and his credit may encompass. It is obvious that the arrival of the gold shipment will be accompanied by a colossal onrush of bills clamouring to be discounted by the Banks of Issue.

Again I ask: how does the movement originate? In money, in wares, in division of labour? Nothing of the kind; but in a piece of news which the public are induced to *credit*. Credit is the creator, the originator. But let me go on quoting. We now come to observe the effect produced on the rate of interest. Gesell wishes to prove that the Banks of Issue take the wrong course, because they misinterpret the signs appearing in the money market:

For some reason or other—our present monetary system knows of a countless variety—prices rise, their advance stimulating the commercial and speculative desire to buy. The merchants seek to raise money to buy with, whereupon the money-lenders advance their terms. The rate of interest goes up. Now the Banks of Issue come into play. And this is their argument. The rate of interest is rising, which is a proof that we are short of money. Our constitution makes it incumbent upon us to meet the fluctuations of the demand for money through the issue of notes. Now is our time to interfere : we must counter this growing demand for money with an increasing supply, we have to bring down the rate of interest by issuing notes.[1]

In the situation here foreshadowed Mr. Keynes would have the Bank "put on dear money," which the Bank is sure to do. Gesell would have the Bank refuse to discount any bills at all. Why these methods are mistaken we shall discuss below. We shall understand the trend of Gesell's argument better if I introduce at this point his statement, which he makes with due emphasis, that "the administration of the currency shall not entertain any relation whatever to borrowers and to the rate of interest." In other words—since borrowing and the rate of interest are phenomena of credit—the money machine is to be run regardless and independent of credit. We have seen that credit, in all circumstances, precedes and creates money. (Gesell also demonstrates how money is destroyed in consequence of the destruction of credit.) How is the money administration to master the forces, if it is not allowed to act in their initial stages ? Of course it may be argued that credit will behave differently, when it is taught not to hope for a rise of prices or to dread the fall of prices. In fact Gesell has carried his argument to its logical conclusion ; the utter annihilation of credit : in his Natural Economic Order the only method of transacting business is by cash payment.

[1] I am not certain whether such exactly is the train of reasoning of the Banks of Issue. The logic of the theory on which they conduct their transactions demands that it should be such. Practically they do encourage and abet speculation during the first stages of the boom, though their practice is diametrically opposed to their theory. According to their theory business is favoured by a low rate of interest ; they wish to favour business, and yet they raise the rate of discount. No contradiction could be more flagrant. If the point is to counteract the rising tendency of the rate of interest—which is supposed to hamper business—why on earth does not the Bank of Issue do the natural thing : keep down its own lending rate ? In America they once had a slogan which may serve as a model for our case : "The way to resume is to resume." So, too, the way to prevent the rate from rising is, not to raise it.

(To be quite accurate I ought to say that he allows advance payments, which is credit *redivivus* and upside down.) Gesell's favourite way of controlling forces is to oppress, to suppress them. And indeed the system which claims to oppress gold out of the currency must be powerful enough to kill credit and to confound interest. The preconceived and fixed idea that money precedes credit and is the originator of interest has seduced one of the shrewdest and boldest thinkers into delusions so flagrant and shocking.

Now let us try to discover where logic would have led Gesell, if he had proceeded on the assumption that money follows credit. Instead of controlling credit through money he would have sought to control money through credit; and instead of taking as his guide the ratio of the supply of and demand for money, i.e. price, he would have steered his course by the ratio of the supply of and the demand for credit, i.e. interest. Instead of saying that the money administration is to entertain no relation with borrowers and the rate of interest, he would have said that it is not to concern itself with the wishes of buyers and sellers. And see how much better this way of reasoning would have fitted in with the main reason for maintaining a stable standard of currency! What need could there be felt for stability if it were not for the fact that *contracts between debtors and creditors are vitiated by fluctuations of the value of money*? None that I can see. It is not as buyers and sellers of wares and services that we suffer injustice and hardship—or enjoy the contrary blessings—when the purchasing power of money shifts, but as creditors or debtors. If we come to grief as buyers or sellers it is our own fault: unwise and listless bargaining. But if we suffer as debtors and creditors we are victims of a contract which cannot be altered by bargaining. My wage contract binds me. I may receive my salary paid out to me in cash every month; but it is not a cash payment, not the outcome of a deal newly bargained. I do not get what I bargained for, if in the meantime the value of money has diminished: I am cheated in my capacity as a creditor. In business it is all credits and debts. The manufacturer, even though he pay cash whatever he buys, buys on credit: on the belief, that is, that he will be able to recover his investment in the sale of the manufactured article. It is as creditors and debtors that we are interested in the standard of currency being stable and reliable.

Gesell himself stresses this point again and again. For my part I have come to understand that the foundation of the currency is not to be looked for in that welling ocean of wares, which is the hunting-ground of buyers and sellers, but in that solider mass which is made up of what we call capital in the narrower sense of the word.

And the phenomena produced in the capital market, the needs of borrowers and investors, the rate of interest is to be nothing to the administration of the currency? They ought to be its paramount care. The movements of the rate of interest are much more significant than the movements of price, and also they precede the latter. It is not the buyers that create the boom, nor the sellers that bring on the slump, but the borrowers and the lenders. Before the merchants can buy, they must borrow; money and credit are their instruments for procuring wares. And when it comes to selling, it is the money lender that forces on action, that initiates the turn of the tide, by calling in his outstanding debts.[1] To forestall the boom, borrowing must

[1] In 1893 a severe economic crisis occurred in the United States, and was also felt in Europe. How was it caused—or rather engineered? for the real cause must have lain deeper than is assumed by the author of the work from which I quote. The trick was done through the throttling of credit by the banks as money-lenders. The following passage is from *High Cost of Living* by T. Cushing Daniel, published by the Monetary Educational Bureau, Washington, D.C., 1912:

"Fortunately, the Panic Circular sent out on the 12th of March, 1893, by this organization, the American Bankers' Association, is at hand, and can be correctly designated as the panic circular of the organized

be discouraged, and the way to achieve this is to *lower* the borrowing rate for money. For, most emphatically be it said : eager borrowing and buying comes from rising rates and prices, while falling rates and prices blunt the edge of desire. To prevent the slump, lending must be encouraged, and the way to achieve this end is to *raise* the lending rate for money, which prompts borrowers to apply for credit, and owners of money to lend confidently. Buying and selling will follow suit.

This is what common logic would seem to suggest. The logical is also the practical. To control money means to control the unruly multitude of prices ; to control credit means to obtain mastery over just one thing, the steadiest in all nature : interest. All the stability that the world of business has so far enjoyed and is capable of is the gift of the natural stability of interest.

money power of the United States, as it was the immediate and only cause of the panic of 1893. It reads as follows :

DEAR SIR,—The interests of national bankers require immediate financial legislation by Congress. Silver, silver certificates, and treasury notes must be retired, and the national bank-notes, upon a gold basis, made the only money. This requires the authorization of $500,000,000 to $1,000,000,000 of new bonds as a basis of circulation. You will at once retire one-third of your circulation and call in one-half of your loans. Be careful to make a money stringency felt among your patrons, especially among influential business men. . . ."

It is quite certain that general conditions must have been such as to invite and favour, nay, necessitate, the insidious stroke of the Bankers' Association. My point in quoting this passage is not to rouse the indignation of the reader, but merely to show that credit is the element of money, and that money can only be reached and influenced through credit.

VI

BELLING THE CAT

At the beginning of the preceding chapter I briefly touched upon the weak points in Mr. Keynes's practical proposals; this matter requires some further elucidation.

The implications of the passage under consideration are as follows: Business is expanding in consequence of invigorated credit. Prices are tending upward. To put a check upon this undesired development it is proposed to reduce the supply (the output) of credit and cash. That seems logical: there is too much credit, therefore diminish it. Mr. Keynes knows of two methods for achieving this end:

(1) Raise the price of the article.
(2) Reduce the quantity offered.

It is Mr. Keynes, the recognized expert, who makes this suggestion. I will add that Silvio Gesell, the great pioneer, who made the same suggestion nearly twenty years before, has never once seemed to question its soundness. How am I to succeed in demonstrating the fallacy in the argument?

You say that a high price and a low supply fit in neatly with each other; they are the Siamese twins of the realm of economics. And so you are satisfied that the experts are right when they combine the two terms. My first objection is that if a high price goes with a low supply, Mr. Keynes's "partly . . . partly" must be wrong for the simple reason that one cannot—because it does not work—do both of two things one of which is the effect of the other. For the high price is the outcome of the low supply, and there is no partly . . . partly admissible. Price and supply react upon each other, and although a high price and low supply do go together, they eternally tend away from each other. Price is the regulator of supply; supply, which determines price, also follows price, so much so that producers experience a sense of calamity when their work brings forth more goods than they have calculated and bargained for. And money is subject to the same laws and forces as any other kind of commodity ; its price must determine its output.

Mr. Keynes proposes to reduce the supply of credit *by* raising its price ; he also proposes to reduce the supply *while* raising its price. I must earnestly beg the reader to note the distinction which I make. Everybody, not excluding those in high places, knows that the way to stimulate the output of any kind of article is to guarantee a paying price for it. When farmers had to be induced to grow more wheat the price of wheat was officially fixed at a safe level, extra high. The same with money : if you raise the price of money, if you enhance its value, if you hold forth prospects of good profits to be made out of the article, you set everybody panting to obtain money, to procure it by hook or by crook, to produce it, to grow it on thistles if that may be. And with all that the Bank of Issue is requested to refrain from producing the article ! Nay more : it is credited with the magic art of making the public surrender the precious treasure from which extra profits are to flow !

However, let us assume that the Bank will refrain and so waive the advantages which it might derive from its monopoly : will that check the creation of credit, of purchasing power ? The Bank has the monopoly for the issue of bank-notes (cash) ; but cash is not the whole. Purchasing power—more correctly : means of exchange—may be created in a great many ways. Money has its rivals and competitors. The most powerful, in our advanced economic system, is the bill of exchange. Statistics will prove that the number of the various forms of bills grows as the rate of interest and discount goes up. Business men, especially the banks, know from experience that the situation created by a rising rate makes credit safe and profitable. Drafts are freely accepted and endorsed, and drafts act like so much cash money. Bills of exchange are means of exchange; when they are issued in greater numbers and for bigger sums, we experience an increase of our circulating medium, with a consequent rise of prices (which is the effect that renders this form of credit safe).

If we are to obtain a comprehensive survey of the situation we must also consider the developments superinduced in the market of goods and real capital.[1] Dear money, in terms of goods, means low prices, and low prices are the means to the end of high prices. Dear loan money is bound to make for dear goods. The output of goods is automatically diminished when the price of money goes up, because the appreciation of money is the same as the depreciation of wares (money gains ground at the expense of wares), and when wares are cheapened production is inevitably discouraged: goods are withheld, their supply is lessened, and the consequent rise of prices is tantamount to a cheapening of money and a reduction of the real value of interest. The measure advocated by our experts to make money dearer will in all circumstances result in the contrary effect; it brings about the very thing which it is intended to forestall: the rise of prices.

Nor will this be wondered at when the situation is once rightly understood. I will state it once more: the supply of credit is to be restricted, because it is supposed to exceed the needs of the moment, and to be forcing up the level

[1] The object of the raising of the rate of discount is to restrict the output of money with the purpose of bringing down prices. The measure, therefore, holds forth a promise that money is going to appreciate at the expense of goods and real capital.

of prices. The supply is excessive : such is the proposition ; otherwise the idea of reducing it, which is held by all the experts practical and theoretical, would be absurd, and the experts would forfeit their claim to this title. There is too much, an over-supply of the article ; I cannot stress this point too strongly. And the remedy proposed is to—raise its price ! Would it not seem that to lower the price is more appropriate to the occasion ?

The problem is important enough to justify my dwelling on it. My solution inverts the approved theory and is likely to stagger the initiated reader ; I have to leave no stone unturned in trying to prove my contention. Suppose the price of loan money (the rate of discount) is reduced in the situation which we are considering : what are we to expect ? The public turns away from an article that is being cheapened ; nobody wishes to make a losing bargain by acquiring it. For it must not be overlooked that at the moment when bank-rate is declared (in our present assumption it is a reduction) we are at a fresh start : money is not yet cheap, it is going to become cheaper. The process of cheapening, as I have already pointed out, takes time, and while it lasts it involves a loss for the handlers of the article. Naturally every person of average intelligence and experience shuns the article so affected, which thereby becomes unsaleable and forthwith ceases to be manufactured. When the rate of discount is lowered, bills of exchange lose their usefulness and are no longer issued, and that is so much purchasing power and credit destroyed. When money is cheapened every one is anxious to put all the money which he finds himself in possession of to some use : he pays cash instead of at three months' date. In the proportion that money is cheapened goods enhance their value, and are therefore produced more eagerly : the increasing output cheapens them, their price declines, which is the effect aimed at.

But we have become involved in a sort of contradiction.

In fact my last sentence contains two statements which are diametrically opposed to each other: goods enhance their value—the price of goods declines. Let me explain. The regulating influence of price takes time. At the start cheap money made dear goods—the dearth of goods stimulated the output of goods, the increasing output brought down prices, and thus the high price of goods has become the immediate cause of the lower price of goods. I have wished to lead up to this point to make the reader realize the absurdity of shifting the bank-rate about, when the aim is to stabilize prices. Any alteration of discount starts a train of adaptation processes, which are translated into a continual fluctuation of prices.

There is another lesson to be drawn from our demonstration. The terms "dear money" and "cheap money" are ambiguous and misleading. They are used, it is true, to designate the lending rate, and apply to loan money exclusively. However, since money is inseparable from goods, it must stand in some kind of relation to them whatever the term "money" is employed for. What does "dear money" signify in terms of goods? Money is exchanged for goods; hence "dear money" corresponds to cheap goods, a low price of goods. But dear loan money goes with dear goods: when the rate of interest is high, the price of goods is also high. If, therefore, we speak of dear money (or cheap) we must keep in mind that it is not the buying money which is dear, but the borrowed money (which is borrowed with a view to purchasing goods that promise to advance in price).

Economic life knows of no states, it only knows of movements. It has nothing static about it, but is all dynamic. Such terms as dear and cheap, high and low, are really meaningless in so far as they are used to designate a state; they acquire significance when we consider them as denoting starting-points or aims, and when we look upon the forces which everlastingly tend to turn cheap into dear and dear into cheap, to lower what is high and to heighten what is low.

To recapitulate our argument: the assumption is that we have to deal with a situation showing an over-supply of credit. The remedy suggested by the approved theory of bank-rate is to raise the discount. My remedy is to reduce the discount—or rather to leave it—and my reasons are: over-supply must be countered by the cheapening of the article, because raising its price—if it could be achieved—will force its output and further increase its supply. The logic of my reasoning is borne out by the record of economic history: rising rates of interest have always gone with rising prices of goods and expansion of credit: *la hausse amène la hausse.*

Now for the other remedy proposed by Mr. Keynes, " the more powerful method, the weapon upon which one would have to depend in the last resort ": the deliberate reduction of the volume of cash supplied by the Bank of England. This remedy is as impracticable as the first, and it is all the more so because it clashes with the first. One cannot reduce the output of an article while its value is being enhanced: it is against nature.

There is so much cash in circulation, with it a correspondent volume of other forms of credit. In order to reduce these quantities you have to withdraw portions from circulation. From circulation? That means from the hold of those who are in possession of the cash and the credit. For these values do not drift about in a no-man's land. They are assets and very highly esteemed by their owners. So the question arises: what is to induce the holders of the money and credit to surrender the article; and what are they to receive in return? Mr. Keynes—also Gesell long before him—suggests that " conversion loan or any other security " shall be offered. I have dealt with this suggestion above. It cannot be carried out because the money market is wholly and strictly opposed to it. I repeat what I have already emphasized : we have to consider money as any other market ware ; its supply is governed by its price. Mr. Keynes proposes to raise its price ; how can he expect its holders to relinquish it while its value is being enhanced ? They will stick to it all the more deliberately ; no one will sell money for conversion loan ; no one will be willing to exchange an asset which is sure to appreciate for one that is going to depreciate.[1] The utter absurdity of the notion becomes glaringly evident from the manner in which it is presented by Silvio Gesell in *Aktive Währungspolitik* (pp. 57–58) :

> The National Money Office, with a view to regulating the circulation of currency, is empowered to issue conversion loan to an unlimited amount, and to reclaim such loan. Through the sale of bonds against legal tender the circulation of money is decreased and the boom averted. Through the purchase of such loan money is added to the circulation, by which the slump is averted. The advantage of this method lies in its simplicity. . . . The control is extremely simple : there are so many currency notes outstanding, and just so many bonds of the National Debt lie in the safe. . . . This method has another advantage which ought to recommend it to the taxpayer : both the sale and the purchase of loan must yield a profit to the Money Office. For the sale (withdrawal of bank-notes) would have to be effected in the periods of rising prices, when securities with a fixed rate of interest naturally depreciate ; and the purchase of loan (issue of currency) would be effected in periods of falling prices, when securities with a fixed rate of interest appreciate. Thus the Money Office would as a rule have to buy conversion loan when its price goes down, and therefore it would buy cheap ; and it would sell the bonds when their price goes up, therefore it would sell dear. Hence every one of these transactions would result in a profit to the Money Office, i.e. to the State Treasury.

---

[1] By the provisions of the scheme this is what one ought to expect : if the Treasury sells securities and buys bank-notes, the increased supply of the former must cheapen them, while the increased demand for the latter must enhance their price. It is shown below that in reality the case is different : securities at a fixed rate of interest—and all State loan is of this description—appreciate and depreciate concurrently with money, wherefore they are not suitable for counterbalancing the fluctuations of the value of money.

The Money Office makes a winning bargain. But that implies that the other party must make a losing bargain. Before the Money Office can operate its fine scheme it has got to find those who are out to ruin their fortunes by a foolish speculation, and we shall enjoy the pleasing spectacle of State officials—who have nothing to gain for themselves—outspeculating the speculators who have their own prosperity at stake.[1]

Mr. Keynes's device is essentially and entirely the same as Gesell's; he has only stated it somewhat less crudely, which saved him from detecting the flaw in it. But we are beginning to understand the practitioners' contempt and the general public's indifference for the fine schemes of the reformers, are we not?

Our case is not yet completely unravelled. The experts propose a measure which clashes with the premises of their argument: they want to raise the price of loan money when they believe that there is too much loan money about, so that it would seem necessary to reduce the quantity. Raising the price would be the method for meeting a situation in which there is too little credit. A high price must be applied—and it applies itself automatically—as a protection of an article which is insufficiently supplied. Might not the method of the practitioners, which is so heartily endorsed by the reformers, have some good foundation? The practitioners do not usually go out of their way to discover a why and wherefore to their proceedings; they allow themselves to be led on by the needs of the hour. Very well: at the moment when Mr. Keynes would have the Bank raise a barrier against borrowing—and Gesell would have the Bank refuse loans altogether—all the banks

[1] In the passage from *Aktive Währungspolitik*, p. 19, quoted above (p. 165), Gesell says specifically that a rising tendency of prices creates the juncture in which speculators will buy "everything with the sole exception of securities with a fixed rate of interest." These securities are the assets which everybody wishes to exchange for more promising objects. It is precisely in this juncture that according to the reformers, one and all, the Money Office ought to sell securities.

are being pestered for loans by eager borrowers. This is a manifestation, not of superabundance, but of shortage. Its implications are that raising the price of an article so furiously sought after—raising it with a view to checking the output of the article—is to put the cart before the horse. Far from diminishing the output of the article, it stimulates it. People hold money too dear, they grab too much of it : then why not take away the edge of their appetite by cheapening the thing ? You want moneys ? See, we give it you extra cheap. However, be moderate, good friend, lest you surfeit. And why so greedy, when you are assured that you can have all you can possibly want to-morrow, and next month, and next year ?

And how, for the matter of that, can we ascertain whether there is too much or too little credit out ? We have no standard to go by. It is all a question of personal impressions and desires. Business is expanding : some say unduly, while others would have it expand more and more vigorously. Furthermore : is it not rather illogical to wish to arrest expansion while the universal cry is for expansion so soon as no expansion is being felt ? Curb down an incipient boom, is Mr. Keynes' advice. By all means, for the boom is the forerunner of the slump which we strive to forestall. But we have got to rise to our feet after having been laid low. It will not do to pin down forces that want to aspire. We must allow them to display themselves. I suggest that stabilization should be the result of a gradual slowing down of the upward movement of recovery from the depression. When the train has been stopped through lack of steam, to set it moving again it is necessary to put on steam, and to let the pressure increase up to the point when the desired speed has been attained. While the demand for credit is strong it cannot be curbed by a higher price ; it is stimulated by it and rendered all the more aggressive, because when a much desired thing is withheld the desire is exasperated and will manifest itself more imperiously and more impetuously. The better method is to erect a limit somewhat ahead. A fixed and immutable rate of discount provides that limit for the creation of credit and cash, while any kind of manipulation of the rate is bound to keep the see-saw going. In order to find an equilibrium the Bank has to place itself at the point of indifference : no ups, no downs in its policy. The Bank's rôle in the game of business is that of the umpire ; it has to stand aloof. Now the Bank cannot perform this function while playing the game. The game has been played so unfairly because there was no umpire. While the Bank is allowed to interfere actively—to manage the currency—there is no safeguard against arbitrary action on its part. Its directors succumb to the influence of the preponderant interests, however pure their intentions may be. The Bank cannot judge fairly whether the amount of credit demanded is legitimate or not ; it is not in a position to feel and know the needs of the business world.

For the sake of completeness I will add the opinions expressed by some more authors of reform schemes. Foster and Catchings's *Money* (published by the Pollak Foundation, Newton, Mass.) is a book which has had a signal success in the United States. These authors believe that Professor Fisher's method would not work promptly enough nor drastically enough in certain possible circumstances. They propose what Gesell proposed nearly twenty years ago, and what Mr. Keynes has come to regard as the more powerful method, namely, that the Government, in a period of rising prices, shall sell in the open market bonds of the National Debt, and buy such bonds when prices go down. They contend that in this way the volume of currency can be diminished and augmented to suit the occasion. By the scheme of Messrs. Foster and Catchings gold is not to be used as a circulating medium. (The review from which

I get the facts does not mention what is to become of the gold of the reserve.)

In the *American Economic Review*, June 1923, was published *The Stabilization of Gold: a Plan* by Carl Snyder. The solution here suggested is to this effect: all the gold which is being hoarded either in the State Treasury or in the Federal Reserve Banks is to be united into one central pile and used to redeem the circulating notes. There is to be only one circulating medium. To regulate the volume of notes in circulation alternative measures are proposed, which may be combined if need be. These measures are the same as those suggested by Mr. Keynes: discount policy and the negotiation of securities. As regards the handling of discount, Mr. Snyder suggests that the rate should be automatically raised (or lowered) by 1 per cent., if the level of prices goes up (or down) 1 per cent., unless it is found more expedient to reduce (or augment) by some 100 million dollars the stock of securities in the safes of the Federal Reserve Banks. Reduce or augment the deposit of securities: what are securities? Can they be made and unmade at will? Drawn from thin air, dissolved into thick air? From what the learned and expert reformers say and reiterate—they just echo each other—it would seem so.

One method for regulating the currency the American and English reformers have not yet discovered, it seems. Gesell's third line of action has recourse to taxation to distribute the means of exchange needed to maintain the level of prices. *Aktive Währungspolitik* (p. 59) points out that neither bank-rate policy nor the negotiation of securities can be relied on to stand the test of every emergency. As a last and irresistible measure Gesell adds this provision:

The National Money Office is empowered to levy or to pay taxes by way of reducing or increasing the volume of currency to suit the requirements of the monetary standard.

And he makes this commentary :

When prices manifest a falling tendency, to strengthen the circulation of money 10 or 30 or 50 per cent. or more of the national taxation is remitted ; the deficiency of revenue arising to the Exchequer is made up for by the issue of an equivalent amount of new banknotes. If the deficiency amounts to 100 million, 100 million will be paid into the Treasury by the Money Office, which sum is added to the revenue. What the people save by the remission of taxes they can spend for purchases, whereby demand is strengthened and prices are steadied. The volume of cash in circulation has been augmented. (The indirect taxes and customs duties must not be affected by this measure, because otherwise the prices of the dutiable wares would fall.)

The chief advantage of this method is that it operates very promptly, since there is no need of any special measures to distribute the new purchasing power among the people. The money which has to be forced into the circulation is already in the pockets of the taxpayers. Instead of being drained to the Exchequer, it returns to market to enliven the demand for wares, and so counteract the falling tendency of prices.

How very ingenious, how plausible ! This taxation method has, indeed, more and more supplanted the other two expedients in Gesell's scheme. In his more recent publications he discards discount policy and the negotiation of securities. (It may be mentioned in this connection that Mr. Keynes advocates a capital levy as a possible remedy in a monetary difficulty, which is also taxation.) We ought to consider the merits of this suggestion. Taxes are a price : the price of public services. All prices form one whole, inseparable, inextricable. Taxes constitute a very considerable item in the cost of living (cost of production). If they are reduced (or augmented) there is so much more (or less) left for other purchases. So it seems. But the semblance deceives. It is with taxes as with interest : they are not paid once or twice a year, but every day ; they are contained in every price that is asked and conceded. If, therefore, taxes are reduced (or increased) an element is subtracted from (or added to) every price. Very cautiously, and very strangely, Gesell in the passage just quoted puts in this reserve: "Indirect taxes and customs duties must not be included in this measure, because otherwise the prices of the dutiable goods would be lessened." This implies that indirect taxes enter into the prices of goods, while direct taxes do not, which is a manifest error. All taxes are treated as an item in the cost of production, and by saying that to reduce indirect taxes would be to reduce prices, Gesell simply affirms that a reduction of taxes must have this effect generally.

We have to look at the matter not only from the purchaser's point of view, but also from the point of view of the seller. The manufacturer of an article may resist the pull on prices while he has to pay full taxes, but yield when his taxes are lessened for him; he may resist the temptation to raise his prices while taxes are moderate, while the raising of his taxes will furnish him a welcome pretext for giving the screw a turn upwards. Gesell again and again stresses the idea that the expectations of the business world and the general public are the decisive factor in the movements of prices. What expectations does the reduction of taxes create? When prices tend to fall it is because the public wants them to fall, and is exasperated that they refuse to fall. Now the easing of the tax-levy will put it into every head that prices ought to yield, that the producers and merchants have no valid grounds for resisting. The consuming public will not rush out to spend what they save on their own taxes; they will wait until the shopkeeper has surrendered his share in the general bounty. There cannot be the slightest doubt but that the measure, if applied in the circumstances here assumed, must produce the contrary effect to the one contemplated.

Suppose now it is applied when prices are tending to rise: lower taxes when Gesell would have them raised. The consuming public are opposed to this price movement, and at strain for some good reason for resisting it. The reduction of taxes will furnish this reason: every buyer argues that the producer, who has had the cost of production reduced for him, has no justification in further raising his prices; he expects that the rising movement will be stopped, and so he defers his purchases, which means a slower circulation of his money and a curb on prices.

Let these few hints suffice. All things have two sides: a deceptive one which allures or scares the superficial spectator, and a true one which, though it does not obtrude itself, yet will be manifested to him who stops to look, and which neither allures nor scares, but rather inspires confidence while imposing the necessity to face realities. There are, in matters of currency, innumerable ways for going wrong and defeating one's own ends. Generally speaking I would say that the soundness of a scheme is in an inverse ratio to the number and the violence of the means to its end. Gesell's three "strong weapons for the preservation of the standard" ought to rouse even stronger suspicions than Mr. Keynes's "partly . . . partly," which is bad enough. Compare with these proposals my scheme of the interest standard. It says: hold on, don't take action. This makes an end of expectations, and where these are eliminated speculation is foiled, and what is to move prices then?

The reformers are driven to their shifts because they have deprived themselves of the help of the natural stabilizer: gold. I will here emphasize my contention that without gold stabilization must be impossible for many generations yet. My scheme, although it gives first rank to interest, assigns a part hardly less important to gold. While interest has a purely passive function—just keeping still—the active part is with gold. When I said that the way to escape from mistakes in managing the currency was to refrain from action, I left half of the truth unexpressed: inaction forces on action in another field, and an active element is indispensable. But action is immune from mistakes only when it has attained to the perfection of automatism. In my scheme gold is made to act, but its action is altogether automatic, with no trace of conscious management or control.

All the methods for regulating the flow of cash which we have been discussing are based on a faulty conception of the nature and function of credit. Credit is the creator and the life-breath of money. There cannot be more, nor less, money in circulation than credit allows. The number of bank-notes and of the ciphers on them is irrelevant; it is the condition of credit that bestows on them a greater or a less degree of efficiency. Credit, unfortunately, is like the ancient god Janus : it has a double face, and lies to him who looks only at one of its faces. When is credit strong and active, when is it weak and relaxed ? It is weak and strong, active and inactive at the same time, only in opposite directions. This dualism is most aptly expressed by the fact that in the fluctuations of economic levels the credit of one set of values rises while the credit of another set declines : when the price of industrial stocks and shares goes up, the price of securities at a fixed rate of interest goes down—the latter losing the credit which the former gain. Now in the discussion of the problem of the trade cycle this fundamental fact is commonly overlooked. Credit is represented as weak or as strong, without any distinction. It is generally stated and universally believed that credit is strong in times of prosperity, that a low rate of interest is an expression of a good and healthy condition of credit. Now I maintain that a falling rate of interest is an infallible sign that credit—commercial credit, that is—is on the decline and weak. Interest and credit are, if rightly understood, the same thing, so that the lessening of one must also lessen the other. Both interest and credit are hostile to prosperity. A well-stocked country where output has been vigorous and goods are plentiful does not make for a healthy condition of credit; the expectations of all concerned go in the direction of a fall of interest and prices, and who would give or take credit, borrow or lend, on such an anticipation? Before credit can be rightly and successfully controlled, this point has to be grasped, and its implications must be applied.

Another point which has been most persistently missed or misunderstood is that credit is created by the borrower rather than the lender (the banks). The borrower is the active partner in the concern. When he applies for a loan he does so on the belief that conditions will develop so as to enable him to meet his liabilities: he credits the money machine with the capacity to furnish him with the cash necessary to redeem his pledge. He gives credit—if not to the bank direct, then at least to the community as a whole. Without stopping to elucidate this conception in a more detailed manner I would point out its main implication: namely, that in order to regulate credit it is advisable to consider the attitude of borrowers rather than that of lenders. What is it that borrowers strive for, and what do they shun? They shun the fall of prices, they are all for an aspiring tendency of the economic forces. And they are eager for action. Now it becomes impossible for them to act so soon as anything—prices proper, interest (the price of loan money), taxes (the price of public services)—begins to give way, while it becomes impossible for them not to act so soon as anything begins to move upwards. But their inaction produces stronger and more immediate effects than their action: it is a check on commercial credit. Obviously the way to relax the tension of credit, its price-forming activity, is to relax something that it depends on: let down the rate of interest, the rate of taxation, the value of stocks and shares; and the way to put credit on its mettle is to tighten all these prices.

Let me also touch upon one at least of the implications of the former point: the relative strength of credit. Let us suppose that credit, in its industrial aspect, is in its flood-tide. Expansion is the keynote of the juncture. It seems as if there were too little of everything—simply because the phantom shape of credit looms so large : too little "capital," too little of marketable goods, too little labour in the market, and—too little money. I have said that credit and money are one and the same thing, and my present statement somewhat contradicts this view : credit and money appear as separate entities, though closely related. Money is an aspect of credit, not the whole of credit. The volume of money cannot expand so promptly as the immaterial essence credit does ; besides, money loses in substance what it gains in bulk : the rise of prices keeps subtracting from the weight of money through the phenomenon termed depreciation. The creation of currency cannot keep pace with credit while credit continues to expand.

It is in this situation that the reformers wish to have the quantity of money consciously and artificially diminished. Gesell the radical would refuse the issue of new currency ; he would withdraw currency from circulation ; he would sell bonds of the National Debt in order to reduce the quantity of money in the market. In a word, he proposes to do everything to which the needs of the business world are *opposed*. By my interpretation of the problem, all these proceedings would have to be reversed. The quantity of money, we found, is insufficient in proportion to the strength of credit ; why diminish an insufficient quantity ? The remedy surely must be the other way about : restore the right proportion of money to credit, either by augmenting the supply of money or by curbing credit. Money cannot keep pace with credit while credit shifts, we have said ; therefore the only method is to check credit. But this cannot be done either, because credit, taken as a whole, is not susceptible to increase or decrease. What can be done is *to make credit take another turn*. It is excessive in one direction, and insufficient in another direction ; securities with a fixed rate of interest, for instance, are out of credit when commercial values have all its favour, and it is this unequal distribution of its power that causes the disturbance of the equilibrium. The way to remedy this discrepancy is to restore the credit of securities, which can be done in one way only : by diminishing their supply—the Bank must buy securities, create a demand for them. In doing so it not only circulates additional currency—of which there is a shortage—it also deflects capital from commercial to national securities : it reduces the credit of the former, which is excessive, and so curbs the main force making for the inflation of prices.

The same holds good with regard to taxes, which, if I were to advocate a "managed currency," I would have reduced when the point is to arrest expansion. The taxpayers' supply of cash must not be drained while they strain for an increase. When all prices tend upwards, make a breach in the chain of prices, and the trend must alter its course. How is credit affected by this measure ? While the State raises its taxes its credit suffers ; but we restore its credit by easing taxation, and as much credit as is added to the State is detracted from commerce. (The credit of the State is also enhanced when State bonds are redeemed.)

But the most powerful brake on the wheels of credit is the reduction of the rate of interest, this price of prices. When it is proclaimed all the makers of credit—the applicants for loans—lose their bearings and are forced to fall back. If the rate of interest proper cannot be reduced by an official decree, the rate of discount of the national Bank of Issue can ; all the reform schemes are based on the assumption that the Bank must be empowered by law to manipulate the rate of discount freely. (Again, however, I must remind the reader that I am far from recommending manipulation.)

What shall be the crucial test of the soundness of my solution? One of the most noteworthy in the long list of American publications on the problem of stabilization is undoubtedly *Business Cycles and Unemployment, Report and Recommendations of the President's Conference on Unemployment* (New York, 1923). One of the chief remedies here suggested against the dangers of the trade cycle is to the effect that the authorities, both national and local, in charge of the public utility undertakings should lessen their expenditure in periods of inflation and spend more in periods of depression. This is obvious enough. Now all the currency reform proposals which I have been criticizing run counter to this most natural of remedies, a fact which in itself suffices to disable them, whereas my solution is approved by the fact that it fits in very closely with the suggestion.

By my interpretation of the case the State remits taxes and moreover redeems part of its debt when business is expanding. The State, that is to say, is depriving itself of the means to do business and so compete with private enterprise; it has no money available for expenditure on public utilities, and it naturally ceases to assist in forcing up prices by creating a demand for raw material, machinery, labour, capital. When everybody rushes into new debts the State does the proper thing: it reacts against this unwholesome practice by reducing its liabilities. Vice versâ, when the period of depression comes on and threatens to cause unemployment, the State will contract new debts by issuing securities. By this means the excess of currency, which otherwise finds no employment, is drained to the Treasury, and the State at the same time acquires the funds necessary to carry out public utility enterprises and so provide employment. This is the only proper and workable way to set money circulating: the purchase of raw material, goods, machinery, labour, services.

The State and local bodies also levy additional taxes. The tax-payers are thereby warned that if they shrink from venturing their money in useful enterprises their money will be taken away from them, to be set at work by the State and its utility departments. This seeming harshness is perfectly legitimate and appropriate. People who do not employ the money supplied to them by a public institution deserve to be penalized—and they need to be prodded for their own salvation. As to the sale of securities, this is the juncture in which it can be effected, because the investing public will readily respond to an opportunity for placing its surplus cash at a safe and attractive rate of interest. Credit is destroyed by plethora and restored by depletion. Well, the moment when the crisis declares itself is one of general plethora; but the congestion is relieved when the State consents to receive the excess: of money, by the sale of securities; of goods, by placing orders for public utility undertakings; of labour, by starting and operating well-prepared emergency schemes.[1]

A "managed currency" needs to be circumspect. Every single measure must be tested and weighed with regard to its immediate and its remoter consequences. How poorly the schemes so far evolved meet the requirements of an all-round appropriateness is most forcibly evidenced by the above examination, which proves them to be unworkable because they clash with practical needs. If practically unworkable they must be logically unsound. It is time that they should be revised. "A managed stability" is a logical monstrosity on the face of it. Now for the discovery of the natural and unmanaged stability.

[1] If the State is to interfere at all, it ought to apply the physical law of interference. The aim of its action is the smoothing down of the fluctuations, the economic tides, which can be achieved by making an undulating valley (depression) from one direction coincide with a hill (boom) from the opposite direction. The State must be inactive when private enterprise is feverishly busy, and vice versa; the State must buy when private enterprise shrinks from buying; the State must go into debt when the private business man tries to liquidate his debts, and so on. The proposals of the experts advise interference of the opposite kind: cumulative interference: raise the rate of discount when private business raises its rates; sell securities (contract debts) when private business is avid for debts; raise taxes when the capacities of private enterprise are taxed to the utmost.

VII

". . . WHICH HE CALLS INTEREST"

"If Nature had provided us with a stable standard ready made."
—J. M. KEYNES.

WHAT is interest? Latin *mea interest*: it concerns me, affects me, I have it at heart, I have something at stake. Interest is desire, appetite, greed, lust, love, devotion, attraction, repulsion, curiosity, egoism, altruism, readiness, awakeness; it is need, necessity, *embarras de richesse*, too little, too much; it is the opposite of indifference: it is difference. Here we have come to the very root-meaning of the word: interest is difference. Interest is essentially the same thing as energy. No interest is possible where there is no difference, in exactly the same way as no energy can produce itself without a difference. Ask the physicist whether an atom, or an elementary quantum, will move unless there be some difference? Where there is a difference, there is interest, and there is energy, either to attract or to repel—which is absolutely the same, too, since repulsion is merely the effect of attraction in the opposite direction. But interest is also an expression of underlying likeness, of affinity, nearness. To take an interest is to be ready to concede, to cede, to yield, to give in: to conform. Interest causes people to bend from the rigid line of their course and so come round. It thus brings ends to meet and make a new beginning. It is really and truly the curb that forces everything, all life and all movement, into the form of forms: the circle, which is endlessness within a compass. It is exactly the same as that deficiency which everywhere makes one force triumph over another, and without which no one thing could ever be encompassed, because all things would tend away into space. Says Nietzsche in *Beyond Good and Evil* (No. 80): "A thing which is elucidating itself ceases to concern us." We need the gap between our reach and the thing which we wish to encompass if we are to remain or to become interested. Interest springs up as soon as the gap is discovered or new created. And this gap is a deficiency, the insufficiency of our capacity and power in acquiring, or the excessiveness of our longing and appetite, our superior capacity for desiring.

In the domain of economic activity the notion of interest has crystallized into a specific form. But in order to understand this special manifestation, this arithmetical formula, it is well to understand its universal application. Had Silvio Gesell looked at it from this more general point of view he could not have fallen into the strange delusion that interest is an evil and a mere mistake, to be suppressed by some artificial and arbitrary device. Gesell very rightly accounts for interest as the outcome of the difference between the relative powers of supply and demand, the latter being in excess of the former. But he is sadly unphilosophical in imagining that supply and demand would keep active after the difference had been abolished; i.e. that men would go on producing when interest to do so had entirely disappeared. Production cannot take place unless there is sufficient interest in the object contemplated to induce people to concede the interest demanded and necessary. Interest is the connecting link between consumers and producers, the concession which makes a community out of units. Interest is the essence of communion. It is that by which the community, the whole, holds its members together; it is that which is between the parts (*inter est*), uniting them, making parts out of what otherwise would be *disjecta membra.* It is far more than a mere phenomenon ; it is a metaphysical reality, it is a manifestation of Life itself. In a comprehensive study of the problem of interest I have come to identify interest with Will as conceived by the greatest of philosophers : Schopenhauer. I could not well present my scheme of an interest standard of price without at least foreshadowing some of these remoter aspects of the thing.

A standard is what endures firm and fixed. Since the beginnings of economic record the rate of interest has been found to oscillate, as a violin string oscillates, round about a well-defined average. All things devised and made by men have undergone change in the course of time ; interest has endured unaltered. Interest is not an invention, not a convention, not an arbitrary quantity. It is the expression of a fundamental fact of human nature—of all nature. It is the natural standard of the forces governing human effort. In the words of Mr. Keynes : " The stable standard ready made with which Nature has provided us."

Interest is expressed in terms of a fraction, five-hundredths being the medium value. These 5 per cent. make up the extra charge of energies needed to set things moving and to overcome the natural resistance opposed to our appetites by the parsimony of Nature. Applied to money : interest is the shrinkage or consumption necessary to create the dividing space and tension between the two poles of supply and demand. Purchasing power must be so reduced by continuous shrinkage as to find the desired object far enough removed to make itself coveted and interesting. I have quoted the passage of Professor Soddy's book in which he expresses the opinion that interest is the result of the failure of the State to supply a sufficiency of money. That is so, but with a qualification. What will people do when supplied with their hearts' content of money, of purchasing power ? They will develop a host of new and costly interests, which will make such an onslaught on the money supply as to create a new shortage forthwith. I mean to make it understood that it is interest which causes money to shrink and be consumed, and so to move on in the margin opened between demand and supply. Interest is the condition of money circulating, of any kind of economic activity. In so far as I buy and consume, I pay interest; I satisfy my personal interest in the thing procured. My readiness to buy, which springs from interest, is the measure of my readiness to pay interest, every price acting as a collector of interest—all the interest paid being charged to the prices of goods and services.

Now, it is a well-known psychological fact that people are the less willing to pay tribute the less you demand of them; but especially that they altogether cease to pay when the impossible is demanded of them. There are times when interest assumes impossible dimensions. It becomes impossible to pay the price of money when the rate of interest goes down. For along with it goes the tension of energy, the power that makes things possible. And along with it goes the level of prices, and every per cent. that it sinks has to be added to the rate of interest. The book of Prof. Fisher was the first in which I have found this fact clearly expressed. From a recognition of the fact to a correct application of its import cannot be a hard step to advance. All that is needed to prevent the level of prices from falling and interest from encroaching on the debtor's assets is, therefore, to keep up the bank-rate, so as to encourage debtors to stand their ground and hold their own: to preserve the proper tension of energy.

The question as to the right figure for the rate of discount to be fixed at need not detain us very long. To begin with, take the average within the period over which records are available. It may be readjusted if experience should prove it to be too high or too low. The way to ascertain that will be by observation of the prevailing tendency of the general rate of interest. If the latter tends to go above

13

the bank-rate it will be a sign that bank-rate should be lowered—the tension of energy being too high. The rate of interest will exceed its natural bounds when the pulse of economic life, which is the circulation of money, is too quick. Reduce the pressure on money, i.e. the degree of its shrinkage through interest, and consumption will slow down, with it production.

All further details will follow automatically. The currency regulations will be contained in a very few clauses, which I will sketch just for curiosity's sake. The law will provide :

(1) That the Bank of Issue shall discount bills, subject to the ordinary cautions, without any limit or discrimination, at an unalterable rate of discount of . . .

(2) That the Bank shall buy and sell gold at current price as offered or demanded, against legal tender without any regard to the amount of paper money outstanding.

(3) That the Bank shall provide for circulation as many gold coins as may be called for.

(4) That the Treasury shall not be supplied any notes out of the printing press.

(5) That the Bank's net profits, after payment of a dividend not inferior to the average dividend on stocks, shall be paid into the national Treasury.

This last provision seems to me to be necessary to remove the policy of the Central Bank—which I would prefer not to be a State institution—from any temptation to profit-making. The Bank must be in a position to act with perfect independence and, if need be, to forgo profits. It would cease to make an income when, owing to a reflux of notes, its opportunity to lend money fails. Such an eventuality, to be sure, is not likely to occur ; still, it ought to be provided against.

Apart from this the State need not have any hand in the business. And it is best that it should not. It is the State's business to rule, not to transact business. Finance ministers are political personages and pledged to party creeds; their influence on the Bank, if allowed any, is bound to be a partial one. In so far as a supervising authority is found to be necessary it ought to be constituted by the free organizations of the economic bodies, such as the Chambers of Commerce and Agriculture, the Co-operative Societies, the Trades Unions, the Savings Banks (as representing the creditors). A heavy responsibility should be laid on the acting managers of the Bank.

We now proceed to an examination of the soundness of the scheme.

Can the level of prices be forced ?—that is the question. The four schemes that have been the subject of my criticism will assist us in collecting the points on which our scrutiny must be directed.

A. Can the Level of Prices be Forced Up ?

I have quoted Silvio Gesell's opinion that "to suit the issue of money to the requirements of the rate of interest would only force up the rate of interest all the more." From the point of view of the interest standard we are led to ask : what, if the rate of interest (discount) is not allowed to be forced up ? Will it be possible for the Bank to issue so much money as to force the level of prices ? In order to find the answer to this question we must agree as to who is benefited by a high, or a rising, rate of interest, whether the creditor or the debtor. For obviously it is on this that will depend the forces brought to bear on the currency. Now I maintain that the debtor is favoured by a high rate, because it is associated with high prices and only obtained, or conceded, when prices go up. Under the interest standard no honest debtor can be forced to concede a higher rate than the discount rate, because he can borrow of the Bank. So if he does concede a higher rate to private lenders it can only be because he expects a rise of prices. And the money-lender can wish to be instrumental in raising the level of prices only while he hopes that prices will come down again; a hope which, if frustrated, will die and cease from seducing. But where can the cash necessary to raise the price level be derived from, if borrowers do not procure new cash at the Bank, which is alone entitled to issue notes? No increase of the volume issued is, therefore, possible, and the danger of a rise of prices is averted.

Silvio Gesell has objected to me that a ring of speculators may contrive to raise money on bills illegitimately drawn. My answer is that attempts of this kind could succeed on such an insignificant scale only that their effect would not be felt. The gold in the currency will be more than sufficient to undo the machinations of the profiteers. I have dealt with this mechanism fully enough, and will add only two fresh points. Society has a very constant structure, or mixture, of elements. The profiteering speculator is a constant quantity, though it may temporarily encroach somewhat (in times of general upheaval), as the number of sick people in periods of epidemic encroaches upon the number of those in good health. No human contrivance will ever be so cunningly devised as to provide for the normal and the abnormal equally well. Thus the profiteer, who is always with us, and no doubt a very necessary ingredient in the economic salad, need not be specially provided against; such provision would rather mar than mend the scheme. As to gold, there is this to be said: it is the very thing to foil the wiles of the profiteer. Every owner of every little bit of gold has the profiteer in him. Now individual profiteering can be successfully met only through general profiteering. The would-be speculators with the Bank's paper money—a small fraction of the population—will run up against this solid mass of profiteers, the owners of gold. In proportion as an excess of paper money is forced into the circulation, the gold sinks below the surface and ceases to buy (the "gold bugs" come into play), so that as much new purchasing power as appears in the market, just so much old will disappear. Gold is credit incarnate, all its preferences resting on the belief that, whatever happens, gold will come out of the ordeal unscathed. When gold quits the market, it is, therefore, credit that quits; in creating an undue amount of paper credit the Bank simply destroys an equal amount of gold credit. The combine of speculators may succeed in forcing up the price of some one article—and it is not the business of the currency to check such action—but gold, in changing itself into an element of supply, will force down the price of some other article or articles and so compensate the price movement. Thus it seems beyond question that the Bank need not exercise any special vigilance, nor discriminate among those who apply for its credit. The gold in the currency is the vigilant policeman who will put things straight.

However, after having admitted that borrowers may attempt consciously to raise the level of prices by overdrawing on the Bank's credit facilities, it will be necessary to show what means are at the hands of creditors to protect themselves. I maintain, with Silvio Gesell and against Professor Soddy, that too much credit cannot be created unless there is too much cash already out, or scope for an excess of cash to be issued subsequently. No one will give credit if he does not feel assured of the borrower's capacity to refund at the expiration of the term. Now the givers of credit are those who ought to know something about the prospects: namely, the banks. And if the banks do not yet know, it is time they should learn. I believe in men taking some responsibility for their own safety, and would not have the State be playing nurse for those who have the suffrage. People are willing enough to insure themselves against the various forms of damage: fire, theft, hail, shipwreck, ill-health, nay, death itself. But they have not yet been shown how they may insure themselves against depreciation of their savings and investments. Recent experience ought to have made them awake to the need of such insurance. Nothing could be simpler ; here is my scheme.

The banks are the trustees of the owners of money. Now let the Savings Banks introduce into their terms a clause to the effect that deposits will be repaid—and loans reclaimed—not simply by the sum, but by the Index-number, so that if, in the interim between the paying in and the withdrawal of the deposit—or between the giving and the repayment of the loan—the Index has gone up, say 5 per cent., 5 per cent. will be added to or deducted from the amount. What is repaid, then—either by the borrower or to the lender of the Bank—would not be an equal sum, but an equal amount of purchasing power. If I thus have recourse to the Index computation, which I have discarded in another connection as unpractical and ineffective, I beg leave to point out that the case is different. Here the Index is not intended or used to regulate the currency, but rather to check and to test it. Index computation will be a purely auxiliary measure—and an unofficial one—to be applied only in case of a disturbance. Of course it will also assist in forestalling, not only in righting a disturbance. I would have it considered as a precaution against possible, though surely unlikely, eventualities. It may well be that if the Index-number has been maintained perfectly constant over a number of years, the computations may lose their interest and the Index clause be discontinued. I believe, however, that a new currency law will be gradually established by the round-about way of some such private insurance, which is the best safeguard for both lenders and borrowers.

I will here point out, by way of furnishing further evidence of the soundness of my scheme, that one effect of such a procedure would be the stabilization of the rate of interest.

If the Index-number is raised 10 per cent. within a year, the Index clause will raise the sum of a loan 10 per cent.; instead of £50, the interest (at 5 per cent.) on a nominal sum of £1,000 will be £55, i.e. 5 per cent. of £1,100 instead of 5½ per cent. on £1,000. If the Index-number is reduced by 10 per cent., interest at 5 per cent. on the same sum will be £45, i.e. 5 per cent. on £900 instead of 4½ on £1,000. The rate of interest does not vary, while both lenders and borrowers get and give the same quantity of purchasing power. All the currency reformers are agreed that the expectations of the business world are the main factor among the forces making for fluctuations. There is no great scope left for general fluctuations when extra profits and losses arising from borrowing and lending are excluded, which they are by the method here under discussion. And so the device calculated to right the wrongs super-induced by fluctuations would prevent fluctuations from taking place and wrongs from being inflicted. The republics of borrowers and lenders need not go on suffering injustice until parliaments are aroused to the degree of awareness necessary to pass a new currency law; it is within their own power to establish justice among themselves. Gradually it would be proved, by the observation of events, that the stability of the purchasing power of money is strictly associated with stability of the rate of interest, and it will be discovered that the simplest method to achieve the end is to establish a stable lending rate and be done with superfluous computations.

Suppose now the banks, more particularly the savings banks, have adopted the policy here sketched: what will be their method for curbing the inflationary tendencies which we have suspected in their debtors? They must forestall the undue creation of cash by the discount bank, and this end they can attain by lending their own credit as liberally as possible (keep out the competitor). They must reduce their lending rate. The point is to prevent the issue of new money, to lessen the tension, and so weaken the energies making for a rise of prices. New money will not be called for at 5 per cent., while old money is to be had at 4 per cent. New cash will not be created by the discount banks, nor credit by the commercial banks, while the savings banks are cheapening money, and so creating conditions that will rather bring down prices, to the confusion of reckless borrowers and lenders. Such a policy on the part of the savings banks may seem a crazy idea : fancy taking less interest when more is offered ! I have heard it affirmed that the superior fighting method consists in yielding a point when the opponent waxes aggressive ; the trick has a Japanese name. It is beyond question that the reputation (the credit) of the banks, the people's faith in them, has been badly impaired of late years. The banks will find it expedient, one of these days, to make a special effort to recover the public's belief in their capacity to administer the people's savings. I have no doubt that the banks which will first offer the extra security of the Index-number clause, will take the wind out of the sails of those who are slow to understand the need of the hour.

B. Can the Level of Prices be Forced Down ?

Professor Soddy is concerned lest there be no chance for society to extricate itself out of the toils of debt. Capitalists, creditors, he thinks, need not and will not allow debts to be paid off. It is the same idea that underlies Silvio Gesell's theory of interest : so long as money does not consume itself and the owners of accumulated wealth are in a position—by withholding their means from common use—to interrupt the flow of the economic forces, debts cannot be reduced, and a more equal, and equitable—and generally profitable—distribution of wealth cannot be achieved. There is a misconception lurking at the bottom of this view. Our new theory of interest, and our scheme of an interest standard of price will enable us to detect it.

To begin with: if I am to save part of the wealth which I create and acquire I must find one who is willing to use what I save, and so contract a debt to me. Debt is a necessary condition for individual saving, and it follows that debt has its useful side. Now the mere fact that somebody uses, that is consumes, my saved wealth, the surplus of my production over my consumption, implies that he consumes more than he produces. Therefore he must be consuming his substance, his economies, his investments, and it cannot be true that debts are not repayable.

But we must grapple with the problem more tightly. Investments are continually being consumed. We have seen the accumulated savings of whole nations consumed within a few years through the mere shrinkage of the value of money. It was a most ruthless execution, but it is the lesson to teach us that the shrinkage of the value of money is the means to the end of forcing creditors to consume their wealth. If creditors neglect to consume it, their debtors will, because consumed it must be. In the Germany of the inflation period there was a great deal of high living and those who indulged in it were all debtors. What they spent in low pleasures and luxuries was the wealth of the creditors. The money of the interest standard currency will have stood the test when it is proved to be shrinking money. All that has to be established is that creditors will not be in a position to force the level of prices down, in the same way as we have proved that debtors are unable to force it up.

The way for them to proceed would be to withdraw their money, or credit. The effect of concerted action in this direction would be to bring down prices, which is tantamount to a check on consumption and to the appreciation of money claims, i.e. the general swelling of debts. The connection between debt and consumption is here clearly brought out. If creditors are debarred from withdrawing their credit they cannot escape the necessity of consuming their accumulated wealth. Suppose the attempt is made: debtors are given notice to discharge their engagements; would-be borrowers are turned down. Debtors finding their credit and means to carry on business failing will have to take action. They apply to the Bank of Issue to discount their bills of exchange, which the Bank cannot refuse to do. Thus the credit withheld by the money-lenders is re-created without any loss of time by the Bank of Issue—or rather by the borrowers acting in self-defence. Moreover, gold will come to the rescue of debtors. A very striking instance of how gold turns out in periods of renascent credit was produced in Switzerland.[1] A year after the outbreak of the crisis the country saw the gold coins, which had been under cover for eight years, reappear in great force. They reappeared in spite of the fact that in no other country gold was allowed to circulate. Under the interest standard the conditions for gold coming out will be vastly more favourable, and it will not be a year, but barely a month before its forces are brought into play. What with new notes procured by the holders of marketable goods, and what with gold beginning to circulate more abundantly, the pressure of failing credit will not take effect, the level of prices will not flinch, confidence remains unshaken, the crisis will not set in.

We have yet another string to our bow. The savings banks, the trustees of those capitalists who usually come to grief in a crisis, are by the Index clause of their contracts forced to join in the efforts to maintain the level of prices. They must not cheapen the money in their trust, i.e. lower the rate of interest; but they have to warn their depositors that unless their moneys are properly employed they will have to make deductions on repayments: the deposit sum is automatically reduced by the fall of the Index-number,

[1] I had written "failing credit"; but when gold takes heart the tide must already have turned, although it is still at its lowest. With a stable rate of discount credit is unfailing, and any incipient fluctuations will only be felt by the most sensitive of instruments, which is gold.

which naturally also reduces the money interest. Would-be borrowers and present debtors, on the other hand, are encouraged by the prospect that, if prices should fall, their debt will be thereby reduced and loss of substance averted from them. So debtors will not be in a hurry to realize their stocks, creditors will think twice before they call in their outstanding debts. Everything considered there is no possibility that a slump should be produced. Debtors as opposed to creditors have always been the stronger, the bolder, and the more aggressive body. With the currency law favouring them, with gold being their ally, with an improved method of banking, the position of debtors will be strengthened so much as to warn even the most powerful and determined money combines from attempts at breaking the standard.

For consider the consequences if the gap caused by the desertion of their capital is immediately filled. These laid-up credits constitute a reserve of purchasing power, an excess of it, and purchasing power is thereby menaced. When, after a lapse of time, it appears that nothing is to be gained by holding out, and the reserves are brought out again, it may well be that prices will rise. The balance-sheet of the speculators will then show a loss of interest on the capital withheld, and a further loss from impaired value. The money of the "bears" has suffered a shrinkage. It is not likely that the lesson will need to be rehearsed very often.

Our money, then, has the true quality of waning money. It is perpetually on the downward slope, the ground being drawn away from under its foothold; it is unable to stop and take breath; it is like the jade of Mr. Pickwick's cabman: forced to move if it is to escape from being run over by the forces which it has set moving. And in order to keep its path free it must clear away as much old capital as it creates new. The owners of this capital have no other choice but to consume it or let it be consumed either by others or by the worms of decay. Not the suppression of interest, called usury, is the problem to be solved by us, but the freer life and display of the forces of interest. Interest languishes when capital is allowed to accumulate through under-consumption. Our civilization will take a fresh lease of life if sufficient consumption removes the obstacle in the circulation of money. And our new currency will see to that. It will keep on building houses when landlords would rather have it rest a bit. The families of the builders will quit their tenements and move into the new homes. Their interest will be transferred from the former to the latter, and where interest gives notice, capital ceases to exist. The tenement capitalists will have to be quick in eating up their property. They can do that by charging, what they receive as rent, to the account of wear and tear. Instead of living on their rent revenue they will be living on their principal. The new capital uninterruptedly produced will push the old capital off the track in much the same way as young buds push old leaves off the oaken spray. Suppose the landlord persisted in not consuming his asset : he would suddenly find his house unrentable, unsaleable, and whatever the value he had given it in his ledger, such value would be annulled. The argument here set forth should be compared to Silvio Gesell's as presented above (p. 99). Gesell, too, assumes that production will be continued against the wishes of the owners of real capital. But he overlooks the fact that new capital destroys old capital and reasons on the assumption that the new capital will be affected like the old. The rate of interest, he thinks, will sink all along the line, for up-to-date desirable houses and industrial plant no less than for undesirable objects. That is obviously a delusion. Life does not proceed on such lines. To bring about the necessary shrinkage, it does not diminish the vitality of all; it kills some individuals outright that the others may live all the more freely. The same method is employed by Life with capital : it kills

Capital—it takes itself away out of Capital—that Capital may live. Did we not say above that interest was the principle of Life, Life itself ?

Debts have to be consumed and renewed. The two main economic and political problems of the world at present are : who is to consume the Reparations Debt, and who the excess of the American gold reserve ? It is even now plainly seen that the French and Belgian people can live and prosper on their own work and production : then who is to consume the Reparations wealth ? Obviously the recipients of it will have to desist from production and devote themselves to consumption. However, it is a fact of nature that he who consumes without producing consumes himself and is making for his end. And so the creditors of the Reparations Debt are literally threatened with extinction, from which they can save themselves only by deliberate destruction of the excess of affluence. The other problem, that of the American gold hoard, will be dealt with in a special section of this book.

Professor Soddy's concern is very legitimate so long as we labour under uncertainty as to how money will behave. Under the present currency system the consumption of debts can at any moment be stopped and the burden of debt made heavier through appreciation. For a number of years I have been trying to make it understood that the remedy for the ills of our time is an increase of consumption. In 1921, when the first English Labour Chancellor of the Exchequer, Mr. Philip Snowden, was still venting his superior knowledge of economic matters in the *Labour Leader* by advocating, as lustily as the best of the champions of the accretion of debt, a policy of deflation, I took the liberty of writing to him to point out why I thought it more advisable to force on more vigorous consumption. Mr. Snowden's achievements in office suffice to make it clear that his superior knowledge was beyond and above my reasons.

Professor Soddy is moving on the same lines as I. He, too, believes that society does not consume enough. When socialists and their enemies were unanimous in advocating economy, Professor Soddy was reported as having said in a lecture that what was needed to end the crisis and unemployment was "an orgy of spending." In his book he writes to the same effect, and I consider his insistence on the necessity of more liberal spending as the superior merit of his contribution. One short passage I must here quote, because I wish to emphasize a difference of my point of view (p. 26):

> Just as people suppose the world is still poor, an equally common political mistake made is that useful and profitable occupation can be provided for all apart from a generally much more developed and ambitious public life.

I take this to mean that the State should be entrusted with the spending of the wealth created by the people. The State can do something, to be sure. But the State, if given the means, will have armies with cannons, flags, and trumpets; and the State will corrupt its servants and citizens. I would advise that the people should learn to spend their own wealth decently. There are plenty of ways, and the ways will be found when it is understood that the world is no longer poor, and that to live opulently is the wisest course we can take. Either opulence with works of peace, or truculence with warlike swagger and aggression. The physical energies set loose by science will produce wealth, which, if left unconsumed, will explode us.

There is no menace to the wealthy in the views here set forth. The wealthy are simply warned that they ought to consume their own, for safety's sake, and not to hinder the creation of wealth for those who labour. Although under a stable currency standard the cost of living is not supposed to rise, there is this to be considered. For the wealthy the cost of living is likely to rise somewhat, which will assist them, very gratefully, I hope, in consuming their excess. The wealthy employ a great deal of service, and the price of service will surely go up. And if the articles of common use, say the necessaries of life, do not become dearer, the articles of luxury will, simply because an increasing number of people will find the good things within their reach. Of course the wealthy will not like it. There is, indeed, very little for the wealthy to like among the prospects of our period. They are much to be pitied. But we, the workers in the field of thought and science, will glory in grand opportunities opening ahead of us.

C. The Scheme is Sound

Gesell makes out a very good case in showing that bank-rate policy, as pursued under the gold standard, was inadequate to produce results. To protect the gold reserve, he points out, it would not have been necessary to raise the rate to prohibitive levels. The moment that merchants began to export gold they knew that the good times were at an end; they would cease to buy, they would have no bills discounted; bank-notes would speedily return to the Bank. That is obvious. However, Gesell does not draw the obvious conclusion. He does not show that the method for counteracting the price movements in the market would be to invert the gold standard practice, i.e. to lower the rate when prices and the rate of interest tend upward, to raise the rate when prices threaten to fall, and the rate of interest declines. He proceeds from a curiously mistaken assumption (p. 57):

In troubled times solvent citizens will not risk their capital; they prefer to keep aloof and so they are not in a position to present any bills for discount. In these circumstances, which are invariably accompanied by a sharp decline of prices, the Money Office would not be able to procure the number of bills of exchange necessary to arrest the slump. The office might, in such cases, reduce the discount for three months' bills to ½ or 0 per cent., and yet not succeed in placing money with solvent debtors.

It is impossible to issue "cheap money," that is the import of this statement, which is, so far, correct. But it is wrong in another respect. Not many pages previously Gesell produces this reminiscence :

I well remember that on the day when the Revolution of 1889 broke out (in Argentina), and while stray bullets from the interior of the town were striking into the walls of my house, my neighbour measured out a building plot opposite and made a contract with a builder for a new house. "What am I to do ?" he said. "I have money, bank-notes, which I am going to protect from being diluted by new issues. I am going to sell it for bricks, which cannot be diluted. . . ." The revolution fell flat in consequence of general indifference, chiefly because everybody, in order to protect their fortunes from dilution, began to build, to sow, to produce.

We all have discovered that prices began to slump, not when times were troubled by war and revolution, but after the return of peace had restored some degree of confidence. Furthermore, it seems truly wonderful that to this day the disciples of Gesell should clamour for the reduction of the rate of discount to arrest the fall of prices. It does not work, Gesell says, and he proves it, too, with very valid reasons. But how slow those who imagine themselves to be knowing are to scent an error in their own systems ! There was once a carpenter who, in trying to fit a plank, exclaimed: "Here I have sawed off two bits, and still the plank is too short !" Traditional bank-rate policy is of a piece with this carpenter's wisdom. It does not work, because it is the wrong way to proceed. In order to prevent the cheapening of things one must not cheapen the thing which is contained in all things : money.

The Bank of Issue, in raising and reducing the rate of discount along with the rising and falling of the rate of interest, made the mistake of trying to run with the hare and hunt with the hounds. Instead of crossing the practices of the money-lenders it imitated them. Let us, then, consider, for the sake of the argument, what the effect of the contrary policy might be. Shylock's chances are spoiled if the Bank issues cheap money when the rate of interest tends to go up. That does not call for proofs. However, the point is to put the break on the circulation of money : will this desired effect be produced ? that is the question. Yes, it will through the uncertainty into which would-be borrowers are thrown ; they hesitate, they bargain with private lenders. And those who, having borrowed, have become debtors will also act : they try to end their contracts, to return dear money to their creditors. Thus there is an immediate slackening of the demand for loan money and consequently, as loan money is money which wishes to buy, there is also a slackening of demand for goods, whereby the rise of prices is averted. Vice versa : in maintaining, or raising, its rate when the rate of interest flags, the Bank warns debtors from hoping any good from a reduction of interest, and from falling into the trap laid for them by the money-lenders. Or : it encourages the money-lenders to stick to their guns, and so force debtors to keep active. The slump has always been hastened on by the disheartened listlessness of the debtors, the only remedy for which is the fillip given by a dose of interest, by the heightening of the tension of energy.

However, an even better course than moving bank-rate in the right direction is not to move it at all. The wisest and shrewdest of thinkers and political economists, the Chinese Laotse, says : " He who acts undoes it." The Bank must not act, if business is to act rightly. When perfectly stable and unaffected, indifferent, bank-rate becomes the safe fortress of the party threatened by defeat —through its own folly. When prices tend to rise, creditors are prevented from falling a victim to their own greed :

they cannot raise the rate of interest and thereby stimulate borrowers to buy and force up prices ; they are rather induced to moderate their claims, which damps the ardour of borrowers, these would-be speculative buyers and profiteers. It relaxes the tension of economic energy. Again : when prices tend to fall, debtors are prevented from succumbing to their own fears : knowing that money is not going to be cheapened to later borrowers, they need not dread the competition of the interloper ; they are not seduced into expecting deceptive advantages from a lowering of the rate of interest, but are encouraged to hold their own.

I will here add a remark on another inconsistency in Gesell's argument. He strongly emphasizes the need of rousing the public into some degree of alertness and understanding in matters of currency. The public is to be made aware of the importance of it by the measures employed to keep it stable. He advocates that the issue and withdrawal of purchasing power shall be managed by means of taxation : taxes to be reduced by 10 or 20 or 50 per cent. by way of distributing additional purchasing power (when prices tend to decline) ; taxes to be augmented when it is necessary to curtail purchasing power. He writes (p. 60) :

> This method, besides being immediately effective, has the very great advantage of stimulating the people's interest in the currency. . . . The irritation will induce the public to try and understand the currency, and this public understanding will contribute to the stability of the currency. Now the best safeguard against the abuses practised hitherto by the banks of issue is to be found in the general awareness of the people. The true protection of the currency from being abused by the State does not reside in the metal, the money material, but in the knowing vigilance of the people.

Most excellently conceived. However, the measures proposed by Gesell are of a nature to relieve the people of all cares. In no way are creditors and debtors called upon to keep a sharp look out and defend their own interests. A State-created office is to do all the spying and watching and warding, while the individual is to repose in full placidity. Gesell believes in division of labour. The labour performed by his Money Office, all the planning and measuring, the weighing of possibilities, the computing of requirements, the publication of statistics : will reduce the private individual's responsibility for his own success and welfare. Commerce made easy : and indeed, Gesell's Natural Economic Order prides itself on bringing commerce down to the level of the lowest kind of occupation. I should think that the law of compensation will make some other thing difficult which is easy now. An "active currency policy" supposes inactive creditors and debtors ; what is performed by the State need not, and cannot, be performed by the citizens. Either the citizens will rot away in slothful inaction, or they will develop their speculative and acquisitive propensities to a point that their new tricks and shifts will undo all the vigilance of all the State's Money Offices. I would rather have creditors and debtors worried into some degree of alertness by the inaction of the money administration. Let every one be sure that the policy of the Bank of Issue will not vary one jot, and will be absolutely neutral, never favouring either one side or the other ; arm creditors and debtors with good weapons of self-defence : and leave the rest to happen as it may. Up to now the weapons for self-defence were unreliable, because they were not properly understood and applied. All that is needed is the knowledge that interest, as manifested in the rate of discount, is the governing factor in the currency. But an "active currency policy," a "managed currency" is the undoing of the people, the very worst and most corrupting form of tutoring by the State.

VIII

THE INTEREST STANDARD AND THE EXCHANGES

Mr. Keynes has made a masterly analysis of the reasons why stabilizing the internal price level is vastly more important than stabilization of the exchanges. But nobody would care to say that the latter is not worth being attempted also. We are now going to see whether the interest standard of currency is able to achieve this end.

Gold achieved it in the past; gold, which the interest standard retains, will do it in the future. It is the natural leveller. But interest also has always been a powerful leveller. Mr. Keynes testifies to this when he says (p. 160):

> Moreover, the movement of the rate of interest up or down sometimes had more effect in attracting foreign capital or encouraging investment abroad than in influencing home prices.

Home prices were not influenced because the flow of capital to or from the country levelled out inequalities; that is to say that home prices *were influenced*, influenced in the direction of stability.

By combining the two levelling agents we are likely to obtain good results.

To show the difference between the pre-war method and the system here advocated, I again make use of Mr. Keynes's contribution (p. 159):

> If gold flowed out of the country's central reserves, this modified discount policy and the creation of credit, thus affecting the demand for, and hence the price of, the class of goods most sensitive to the ease of credit, and gradually, through the price of these goods,

spreading its influence to the prices of goods generally, including those which enter into international trade, until at the new level of price foreign goods began to look dear at home and domestic goods cheap abroad, and the adverse balance was redressed. But this process might take months to work itself out.

The modification of discount policy, we are told, was the decisive influence in bringing about a new price level. I have already quoted Silvio Gesell's opinion that it was unnecessary to raise the bank-rate, because the mere fact that gold was exported and flowed from the central reserves was sufficient to warn merchants that credit was being restricted. My theory is to the effect that the raising of bank-rate ought not to depress, but rather to raise, price, and does raise it unless it is over-balanced by other influences. I will not resume this subject now ; for happily with a fixed rate of discount we are rid of the vexing problem ; discount policy cannot be altered to suit the movements of gold. Something else will have to adapt itself.

Suppose England to have established an interest standard ; now watch events. We expect stability to favour industry and commerce, and to cause an expansion of all business activities. But we must not expect that exports will develop faster than imports, since the first and fundamental effect should be readier and more regular consumption. Thus we may assume that imports and exports will balance, which cannot but be the normal and healthy condition, nay, inevitable, under a stable standard. It is clear that this circumstance will contribute powerfully to stabilize the exchanges. Indeed, the service asked of the levelling agents is merely so to regulate the flow of values to and from the country as to make them balance. The rates of exchange are an aspect of the ratio between consumption and production. The exchange rate of countries will rise when they over-produce (under-consume) and fall when they over-consume (under-produce). Now to over-produce is to add to the stock of accumulated wealth, to over-consume is to reduce it. The stock of wealth is contained in a country's gold reserve and investments abroad; the wealth in daily use, such as houses, industrial plant, means of transport, etc., however great, cannot be considered as a reserve, and neither can the stock of exportable goods, because they are in the daily current anyway, and cannot be disposed of unless a market is found for them. From this consideration it follows that when gold flows out of a country, such country is drawing on its reserves—which, by the way, is far from being an evil in every case, but may be necessary for the nation's welfare, since too much fat is apt to develop into a menace.

English merchants will begin to export gold when they find that gold buys more than currency paper, because the exchange rate has been impaired. They procure the gold by whatever means they can. There is no fixed price for it; but we have found that gold must keep at a parity with paper money so long as it is used as money. Probably merchants will retain what gold coins circulation brings into their possession; but they may also buy it at the Bank against paper money: it does not make any difference. The export of gold at all events diminishes the volume of money in circulation. The effect of this, of course, is to put a brake on consumption, as it diminishes purchasing power. Thus the cause making for the lowering of the value of money abroad, viz. excessive imports, is eliminated: with consumption curbed, imports are bound to be reduced.

However, there is a hitch in this argument. Merchants may go on importing goods and exporting gold, the outflow of the metal being compensated for by new issues of paper against discounted bills of exchange. But can they? No, they cannot, although no law restrict the drain of gold from the central reserve. A country consuming too liberally —which means high living—is sure to consume its full share of gold: articles of luxury must be in good demand. Therefore the lessening of the gold supply will speedily make itself felt; the price of gold will tend to rise above the parity, and gold ceases to be supplied as currency. The circulation is now further reduced, consumption is further discouraged, and imports must diminish.

Nor is this all. We now have to turn and see how production is affected by these movements. The outflow of gold is superinduced by the push of prices upwards, which encourages imports from cheaper markets. But it also encourages production—imports themselves being but a sort of production. Long before the gold supply begins to give out increasing output will come to the rescue. The flinching of the rate of exchange—which is the cause of the export of gold—facilitates exports, so that foreign bills will be more plentiful and suffice to restore the balance. It may safely be asserted that none of the processes sketched will be carried to appreciable dimensions, because action and reaction will be almost simultaneous. It is hardly necessary to mention that if the sale of a great nation's gold were carried too far, i.e. beyond the danger-line, gold would be so cheapened in the foreign markets as to begin to flow back again.

Mr. Keynes expresses an apprehension that a country may export too much of its gold: "Nowadays the gold reserves might be dangerously depleted before the compensating forces had time to operate" (p. 160). The danger arises, he tells us, when the readjustment of internal prices does not proceed quickly enough. We have shown that gold, if unhampered in its movements, is as prompt in levelling out disturbances internally as it is in preventing external ones. So Mr. Keynes's apprehension may be confidently dismissed.

It may be asked: where is the stability of the rates of exchange if, as has been assumed in the preceding chain of arguments, it does show signs of flinching? Well, stability should not be confused with rigidity. Ours is a natural system, and there is no rigidity of details in nature.

In the past the rates of exchange used to fluctuate between the gold po·nts, and nobody complained. Under our system the gold points are likely to move even closer together. It will be a very excellent self-regulating system, nature itself.

Let us, by way of making the demonstration all the more complete, also consider the opposite case, viz. the accretion of the national stock of wealth, gold flowing in rather than out, in consequence of an excess of exports over imports (under-consumption). So long as the foreign customers of England have gold to pay with, the rate of exchange cannot be affected. But when their supply gives out the sterling will gain in value. That will mean a check on English exports, English prices becoming too high ; it will also force imports, foreign prices coming to be lower than English. The effect on the internal price level will be a depressive one. Now the gold automaton will begin to play for an increase of circulation, which in all circumstances is an increase of consumption. The public circulate their gold, which loses ground if kept as a ware ; consumption grows brisker and assumes healthier dimensions ; prices tend to rise again, and the rate of exchange is restored, the super-valuation of the pound sterling being removed through an increase of imports, a diminution of exports.

It may be that English capitalists will try to evade this issue by exporting capital. This action can only defer, it cannot prevent the return to sanity. For the exported capital will force on the import of goods in payment of interest, and it narrows down the market range of English exports, because the capital exported transplants production to regions which used to import manufactured goods. Thus the situation will be aggravated until the tension of a disturbed equilibrium will turn the flow of forces. Things cannot go too far, under our system, because home consumption is sure to absorb the output by destroying the excess of capital through depreciation. Enough has been said above as to how the new currency will deal with greedy and restive capital. The mass of the working population are going to be the great consuming force, and they will learn how to turn out of office ministers of finance who glory in economizing budgets after the fashion of the first Labour Chancellor of the Exchequer.

IX

THE AMERICAN HOARD OF GOLD

WHAT has been made out so far is that gold in a currency is a very great convenience. But it is not by any means a necessity, so that no nation need consider itself handicapped if without a proper stock of the precious metal—so long as it forbears to live beyond its producing capacity. On the other hand an over-supply of gold may cause very considerable embarrassment, in so far as all those things that are normally considered as an asset will be a source of worry, if that asset cannot be fructified and enjoyed. Gold may turn into mere dross ; but it will be dross with a nasty sting and goad to it that will not allow men to leave it alone.

I have already quoted the opinions of Mr. Keynes and Professor Soddy concerning the American hoard of gold. It may add to the reader's sense of the urgency of monetary reform if I submit my view of the problem and its possible solution.

The difficulty arises from this alternative : (*a*) if the excess gold is employed to buy, it will inflate the American currency and lessen the value of gold generally ; (*b*) if the gold is left buried, it will seem as if a big asset were immobilized, and agitation to have it set afloat will never be silenced. Meanwhile the treasure keeps on increasing year by year, which fact can only be interpreted to mean that certain nations hold the precious thing so cheap as to deplete their stores entirely. Americans are thereby placed at a serious disadvantage ; they are debarred from spending their wealth and so returning the gold to their customers. It is largely their own fault, though ; they are too greedy and produce more than they consume, which is unnatural and unwise, and bound to inflict certain hardships. America is forced to manage the price of gold, "at great expense," which is fairly obvious, seeing that it can be no small undertaking to manage the unmanageable.

Gold, within the last ten years, has been visited with very serious depreciation. It is menaced by worse, and all because it is disabled through unequal distribution. A great many people are affected by this state of things. German reparations are fixed in terms of gold. Thus the would-be beneficiaries of reparations must be interested in a high price of gold. If it were not for this consideration Europe would by this time have definitely broken away from the fascination of the gold standard. As it is, the efforts undertaken to support, more or less artificially and arbitrarily, the value of gold have been one of the main causes of the economic crisis, and they have cost Europe more wealth than all the gold in America is worth.

America will know what course to take when it appears what policy the European countries are going to pursue. If they throw off the obligation to keep a gold reserve, the American gold will come to them freely, Americans being fain to let it go on almost any terms ; if, contrariwise, they allow themselves to be persuaded that they must have a gold standard with a backing to boast of, the gold will refuse to come ; such is the cussedness of the yellow witch. It is more advisable to leave gold to come freely, i.e. to decontrol, but not to disable it. America finds herself in the very awkward situation of the creditor who has much to lose and nothing to gain through trying to drive hard bargains with his debtor. How could she coerce her European debtors into submission and induce them to believe in her gold ? By refusing to buy European goods ?

But to buy them would be the way to get rid of the oppressive surplus of gold; also the way to get her outstanding debts paid. By refusing to supply her surplus production to Europeans? But that is to choke and stew in her own juice.

The right policy for Europeans is to make themselves independent of gold, while at the same time admitting gold without any restrictions: throw off the yoke without hurting the party that holds it. The proper view of the situation for Americans to adopt is to know that their gold is a dangerous excess of accumulated wealth (fat), to be got rid of in one way only: to wit, by spending more than they get—by over-consumption. I shall not here try to prove in detail that the growth of the gold heap is the happy result of prohibition. Such is, however, my conviction. It will not do to cut out a main item of the cost of living, and moreover render the average worker more efficient through abstinence and sobriety, without at the same time increasing consumption in other lines, such as books, art, recreation, leisure, travel. What have American workers profited through abstinence? Their working day is still a heavier one than for many workers in Europe. So then America must revert to a saner policy: she must consume more than she has been doing. This can be achieved in two ways. (1) She can import consumers: reduce the working day to a point at which an increase of the number of hands becomes necessary—free immigration. (Saving means, in all conceivable circumstances, to economize labour, to employ few hands. Increasing consumption, in all circumstances, implies a more liberal use of labour, the employment of many hands. The shortening of working time, connected with the employment of more hands, is tantamount to increasing consumption, the article consumed more liberally being leisure rather than food or clothing and suchlike necessaries.) (2) She can let the workers whom she would rather not have on her own territory produce for her in their native countries—free imports. It will make little difference; but considering the anti-foreign mood of the free Republic the latter course would seem to recommend itself.

The American creditor is in a sense at the mercy of his European debtor. He cannot realize his outstanding debt except by consuming it. For the new truth is asserting itself that the creditor has got to eat up his claim on the debtor that the debtor may work off his liability to the creditor. It seems that to eat up what is owing to him is a bitter pill for the creditor. Much happier he who is allowed to retrench in order to clear himself of a debt. The gold heaps in America, if they are to be set free, have to be converted into consumable and consumptive goods. Merely to loan them away to foolish Germany, who insists on being chained to a gold standard, would be no remedy, since they would yield embarrassing payments of interest. Consumed they have to be; but of course America is free to invite her debtors to join in the feast by remitting part of the debt. It may yet come to the adoption of this course, because a period of high living might degrade the nation. One perfectly safe path is open: it is for America to contribute lavishly to the creation of works of peace and beauty and joy—at home and abroad. Set the forces of consumption afoot in the direction of what we speak of as the higher values. Find out where money might be helpfully applied: in laboratories, in libraries, in the homes of patient seekers after truth; where eye-sores might be replaced by pleasantness; where a degrading sense of penury and hopelessness might be relieved, and men taught to look up. There is infinite scope for profusion—not by way of charity, but of proudly self-confident work. Has it not yet begun to dawn upon the world that enterprise is the greatest and healthiest consumer? Through enterprise America will succeed in getting rid of the excess of fat on her tissue. It will be to get rid of the threatening danger of international tension and friction created through illiberal immigration laws and other restrictions (tariff barriers). A decisive step forward in the settlement of the world will have been achieved when the gold of the world is more regularly distributed among the nations.

American creditors—the virtual owners of the gold treasure—are threatened with inflation. Their only way to escape from this danger is to convert their claims on money into goods and real estate, which inflation does not consume. Buy, in order to spend the treacherous gold. Hitherto an increase of consumption has always been forced on by inflation ; that is the reason why creditors are always clamouring for economy. A stable standard of currency is apt to lay their fears. The Interest Standard will enable Americans to go through the inevitable period of over-consumption without imperilling their monetary value. We will try to sketch the part which interest will play in the process.

The rate of interest is determined by the ratio of production to consumption. With the latter growing beyond the former the rate of interest rises, which tends to force up prices. To prevent this from taking place the owners of money capital will keep interest low by lending at a rate below the rate of discount. This policy must be deliberately pursued in the face of very powerful opposing tendencies. A low rate of interest, when not associated with declining prices, greatly favours the debtors, i.e. those who are not owners of the means of production, the great mass of workers. Their capacity to consume is thereby enhanced, and it is in this way that the necessary course of over-consumption should be inaugurated. The aim to be attained, we are to keep in mind, is to reduce the gold assets of the country. In the last resort these assets are privately owned, so that to reduce them is to reduce the wealth of capitalists ; it is not a public national authority that spends the gold, but private persons. The lower rate of interest, which reduces their revenue, is the means to the end : if the rich men keep up their normal consumption, their expenditure will exceed their revenue, with the result that their principal has to be drawn upon. Assuming that it is decided that the nation as a whole shall spend more than it gets, what section of it could do so but those who have a fund to draw upon, the capitalists ? If they contrive to shift the task of the actual consuming of their excess wealth on to the working classes—not only of their country—it will be greatly to their credit and to their advantage.

The idea surely smacks of extravagance. Capitalists have never yet been known to invite others to assist them in consuming their superfluity. But we do know of cases, well enough authenticated and not too far removed, when more than the superfluity of capitalists, when their whole substance, was consumed, though greatly against their will, by the non-capitalists, i.e. the debtors. American creditors have to decide what proceeding they would prefer, whether the conscious, deliberate, and measured surrender of their excess, or the catastrophic and ungovernable : make a war with the purpose of reducing the intolerable burden of fat (the purpose need not be felt or understood, much less avowed), or consume the excess by deliberately forcing expenditure while keeping the rate of interest at a low level to prevent inflation. Another method, as was hinted at above, would be to remit a part of the claims on foreign nations. But that does not seem natural ; debts are not forgiven among self-respecting people, neither does it benefit the debtor, but it rather humiliates him. Whichever way is preferred, the necessary and desired reduction of the gold reserves—which represent stored-up consumption goods—can be carried through only at the expense of the owners of these reserves, the creditors, the capitalists. These gentlemen and ladies will have to shoulder the heavy burden : to spend, to be lavish and generous, to act up to healthy principles. It is a severe test, I know : relinquish the throttle-hold on the lives and destinies of nations and individuals—human lust of power will not readily consent to it. And there is another grave difficulty about the matter: it requires some real thinking to grasp it, and it necessitates the influence of an economic science that has learned to look facts in the face and call them by their names. So it may prove futile to advise a deliberate and cheerful sacrifice of superfluity as a means to preserve competence. It is much to be feared that the fattened men will wriggle themselves into the hotter waters of a new war. I think I can see the lean Japanese whetting their appetites. And how all the empty gold chests in old Europe are longing for the chance that will come to them—and is bound to come to them—from the bursting of the American pouch!

The plight of the Great Republic brings a word of the ancient Chinese sage Laotse, that shrewdest of sociologists, into my mind. It is to this effect:

> "To try and retain possession of a thing while filling it too full—it is not worth the trouble;
> To want to manage a thing while keeping its edges sharpened—it cannot be persevered in long;
> A room filled with gold and precious stones—no one can protect it."

Gold, by its very nature, will not keep the peace while it is unequally distributed among the nations. Like any force of nature it tends away from the high places to the low. It is a negative quantity which will give the negative value to everything positive. Its creative virtue is only displayed when allowed to meet other negatives: among debtor nations, in poor countries, on neglected soil. We may be quite sure that the gold of the American treasure houses will be let out sooner or later. It may be in a beneficent way; but my greatest concern is lest it come fraught with destruction.

We have to pay a moment's attention to what will take place in the countries receiving the gold which America parts with. They produce and sell more goods than they themselves buy and consume. Wealth, or capital, begins to accumulate; the rate of interest, under normal conditions, would show a tendency to flag. However, the case is not a normal one in so far as the disposal of the surplus production goes on unhampered, thanks to the readiness of the creditors to over-consume (under-produce): unsaleable goods do not choke up the channels. And this is the reason why this is so: the goods exported in excess of those imported are paid for in gold, which gold is laid by. Now it is clear that he who accepts gold in exchange for goods spends the proceeds of his sales and consumes his profit. For to acquire gold and lay it by is to indulge in a luxury, is to consume the superfluous. The effect will be the same as if the countries importing gold were consuming the whole of their own output. So the consequences of over-production are averted: the rate of interest is not brought low, the price level is maintained, the wolf slump is kept from the door.[1]

Again I must pause a moment to account more fully for the development predicted. The debtor nations, now almost depleted of gold, acquire gold from the creditor nations, who have consented to consume more than they produce. The gold which the debtor nations obtain goes to the creation of a "war-chest against emergencies": it is not employed to buy; it is immobilized, rendered useless for the time being—as useless as ever it is at present in the American hoard. It does not enter into the currency; it does not buy goods, neither is it sold as a ware; it is not invested as capital, and does not bear any interest. It is, in fact, dead and as if consumed. We may put it thus:

[1] It used to be thought that the ratio of the gold reserve to the volume of the note issue determines the rate of exchange. There are countless cases to prove that this is a fallacy. Gold may accumulate in the reserve without the rate of exchange going up, and vice versa.

as the creditor nations consent to consume their debtors' surplus production of goods, the debtor nations consent to consume their creditors' surplus possession of gold; gold being a ware like others, the transaction is an even exchange of commodities, so that the rates of exchange are left unaffected.

Gold, of course, is not consumed in the strict sense of the term. But gold that is laid by is always intended for future use or expenditure, which means consumption. There would be no sense at all in creating gold reserves, if it were not with a view to spending it, consuming it, at a future period. So we may reasonably say that gold reserves, both private and public, are for consumption, and the forming of them amounts to consuming them.

A word about the manner in which the inflowing gold can be directed towards the formation of a national stock or reserve. The gold is received in payment of goods exported, and goes first to the sellers of the goods, who probably have no use for it. They employ it to meet their liabilities to the bank which has discounted their bills of exchange. In this way the gold accumulates in the Bank of Issue, where it cannot draw any interest, because by the provisions of the Interest Standard the Bank cannot increase the issue of notes merely to keep pace with the growth of the reserve. In fact the Bank acquires the gold in payment of outstanding debts, which, when cancelled, cease to exist and yield interest. The gold is immobilized, as it ought to be, in exactly the same way as grain intended for seed must be immobilized. Of course there is nothing to prevent private persons from laying by gold, and not a few will do so. But gradually gold will begin to circulate as money. The Bank will find it convenient and expedient to issue gold coins, and so the monetary organism will recover its normal and regular pulse.

The American creditor and customer, we saw, is going to surrender his surplus of gold, which thereby passes into the possession of the seller and former debtor. The result of the transaction is that not only the gold treasure, but also debts and liabilities, are more equally divided. The relation as between debtors and creditors is dissolved, to be succeeded by one of a more agreeable nature. Nations will continue to be dependent on each other ; but they will meet and supply one another's needs on a footing of equality.

Gold is a spoiler of the peace among nations so long as they erect it into their ruler and judge by subjecting their currencies to its caprice and bringing its shortcomings to bear, rather than its virtues. But this same gold will become the peace-maker so soon as it is relieved of its unnatural supremacy and allowed to move freely : to follow its natural bent, which draws it towards negatives. We are going to emancipate ourselves of it, not to cast it out and debase it, but rather to return it to its dignity and nobility, and to obtain of it services all the more precious.

PART IV

POSTSCRIPT

A POSTSCRIPT

I HAD just completed the foregoing chapters when I was made acquainted with *Bankers and Credit* by Mr. Hartley Withers (London, 1923). This book by a whole-hearted champion of the traditional gold standard, though it did not teach me anything really new, was a welcome addition to my arsenal of facts. Its criticism of the various reform schemes differs altogether from mine. Indeed, it makes hardly one point which I should like to endorse fully. If I add this postscript it is not because I wish to support my theory with Mr. Withers' opinions, but because some of the facts supplied by him seem to me specially adapted to illustrate and round off certain sections of my book.

1. The Beauties of Stability and the Modesty of some Stabilizers

Mr. Withers is all for stability and proclaims its necessity with no uncertain voice (pp. 10–11):

> Then the man in the street found out that money has not only to be freely convertible into goods, but must also, to do its work soundly, remain fairly steady in value, and must, if possible, be convertible into the same quantity of goods to-day, and a week hence and a year hence, and a hundred years hence.

Now, what does gold standard stability look like ? Since the passing of the Bank Charter Act in 1844, the English pound sterling was always the same quantity of gold up to the year 1914. In terms of goods, as expressed by the Index-numbers, this monetary standard varied in a most outrageous manner, as any chart of the movements of the Index-numbers will show. Germany introduced the gold standard in 1873 after the system had been tried and improved in various other countries over a considerable period. Yet the Index of wholesale prices shows the following oscillations :

1873	100	1900	72·8
1879	71·1	1903	66·4
1880	80	1907	82·4
1886	58·4	1910	75·2
1891	79·2	1912	89·6
1895	58·4	1913	77·6

To talk of stability in the face of these figures is a mockery. Very wisely Mr. Withers passes over the panics with which the countries blessed with the gold standard have been visited at least once in every decade. However, I shall not rehearse the charges brought against the system by the assailants of the gold standard, which I have come to consider as exaggerated. The gold standard, which first made paper money possible, was surely a great improvement on the previous system. If it did not work satisfactorily it was because, misunderstood and mismanaged as it was, it did not function properly. But to extol its excellences, to clamour for its restoration in its unamended traditional form, is conservatism of an exceptionally virulent kind. Let Mr. Withers and his party make up their minds as to the meaning of the word "stability." If they want to have the gold standard back again, they must demand it on the plea that it favours every kind of pernicious speculation, that it provides a proud position for a small number of gentlemen whom it may pay to flatter, and that it creates special facilities for wiseacre augurs in the temple of Mammon (all of which attractions are not sufficient to stop, or even so much as slow down, the march of events) ; but they must not attempt to capture our adherence on the plea of stability, for that is taking us for more foolish than we are.

2. The Gold Dollar and Goods Dollar Scheme Reconsidered

I was prompted to make another effort to get at the root and core of the gold dollar fallacy, as contained in the Goldsborough Bill, by an authentic interpretation of the scheme which I found quoted from *Proceedings of the Academy of Political Science* (New York, January, 1923) in Mr. Withers's book, p. 248 :

> The mechanism of the Bill provides for the retirement of gold coin by the payment of a small premium, leaving gold certificates, or yellowbacks, in circulation. There will then be a certain number of dollars in gold in the Treasury, either in gold coin or gold bars. Should the general wholesale commodity price level rise, say, 1 per cent. in a given period (two months in the Bill), the amount of gold in a gold dollar is increased arithmetically 1 per cent., which leaves 1 per cent. less of gold dollars in the Treasury, which necessitates the retirement of 1 per cent. of the yellowbacks, which in turn contracts the total volume of potential money and credit by 1 per cent. This process is followed every two months until the price level is down to normal. Should the price level fall below normal, say 1 per cent., the amount of gold in the gold dollar is decreased 1 per cent., which increases the number of gold dollars in the Treasury by 1 per cent., which permits the issuance of 1 per cent. more of gold certificates, which in turn increases the potential money and credit throughout the country by 1 per cent.

This is far too logical to be true. " Should the price level rise, the gold dollar is increased " : but what has the gold dollar to do with the case ? What has it contributed to the rising of the price level ? You do not find in the Bill one word to explain this. Section 2, to be sure, provides that " the approximate fluctuations in the purchasing power of gold shall be computed from the Index-numbers." But this does not furnish any clue to an explanation. No connection between the " fluctuations in the purchasing power of gold " and the " fluctuations of various wholesale prices " becomes apparent. The Index-numbers are computed independently of gold. And what, after all, is this computing of the purchasing power of gold? It cannot be computed, it can only be tested in the conflict of supply and demand: by bargaining in the open market. Gold is not a purely arithmetical, it is a material quantity. Also it is more than a convention. When its price fluctuates it is because certain needs and influences are manifesting themselves. The fluctuations are the result of, and in their turn translate themselves into, movements in the ratio of supply and demand, of production and consumption. The lives of thousands of men are affected by a small change of the price of gold. When its price goes up the production of gold is stimulated, while the output of other wares—which are cheapened in the same proportion as gold is enhanced—is discouraged. If the state of affairs persists for any length of time labour is withdrawn from the production of wares and directed to the production of gold, up to the point when the equilibrium between the price of the gold dollar and the goods dollar would be restored. The parity between the two, which the Goldsborough scheme aims at, is not a fiction to be set up by a trick of computation, but a reality and a natural necessity which will impose itself by means of hard facts.

An alteration of the contents of the gold dollar would only be justified when it is proved that it is gold that has caused the fluctuation of the level of prices. But by the Goldsborough scheme gold is entirely innocent. The proposal reminds me of the ingenious institution of the "whipping boy," once in vogue in the small German courts. The whipping boy (*Prügeljunge*) had to receive, vicariously, the birchings deserved by naughty princes, who must not be punished. The rising of the level of prices is not brought about by gold—in the mechanism of the scheme—nor is it even made visible by anything happening to gold. It is the outcome of an alteration of the ratio of supply and demand of ordinary commodities in the open market, and proved it is by the computation of Index-numbers. By the theory underlying the scheme the withdrawal of currency imposes itself regardless of gold and gold reserves, which can be gathered very comfortably from our authentic interpretation. The computation of Index-numbers is supposed to show a rise of 1 per cent.; the amount of gold in the gold dollar is consequently increased—"arithmetically"—the same 1 per cent., whereupon the same 1 per cent. of yellowbacks is withdrawn and the volume of money is contracted by the same 1 per cent. Surely the gold dollar is gratuitously introduced into this chain, and can be left out without the final result being thereby affected. The Index-number has gone up: by the rules of the primitive quantity theory the volume of money has to be reduced 1 per cent., which is all that is needed to reduce the price level to "normal" (supposing that the theory be sufficient and error-proof): the gold dollar, like the flowers that bloom in the spring, has nothing to do with the case.

However, our purpose is not to ridicule and to defeat the Bill, but rather to learn from it, as the laboratory investigator learns from an unsuccessful experiment. We have to try and discover the fundamental vice of the conception from which the scheme has been evolved.

According to Mr. Goldsborough's interpretation the alteration of the gold dollar and, along with it, of the reserve, is merely "arithmetical," i.e. not a material one. The value of the reserve (or the number of gold dollars in it) is increased or decreased without any gold being added to it or withdrawn from it. Thus it would seem that, as I have already hinted, the quantity of gold in the reserve need never be altered at all. We should then have a kind of gold standard on the basis of a fixed and immutable gold hoard, instead of, as by the traditional gold standard, on the basis of a fixed ratio of the weight unit of gold to the monetary unit. The wording of the Bill itself does everything to make one believe that this is the real aim and intention. The procedure of issuing and withdrawing "gold bullion dollar certificates," with a view to maintaining the reserve at 50 per cent., is described in these terms :

> The Secretary of the Treasury is authorized to make said withdrawals of certificates from circulation by withdrawing from the Government deposits in national banks, and to issue certificates and place them in circulation by adding to those deposits.

It does not provide for the purchase or sale of gold, which would augment or diminish the reserve, but purports to work the adjustment by the manipulation of Government deposits of paper notes called "gold bullion" certificates. Surely the withdrawal of gold bullion certificates ought to be an exchange, dollar for dollar, of certificates for gold : the reserve to let out as much gold as it takes back in certificates or lawful money. This, however, is proved not to be the intention of the scheme by the fact that the reserve is to contain, in gold, only 50 per cent. of the value of "the gold bullion certificates outstanding from time to time." On this provision the exchange of gold for certificates cannot be effected.

On the other hand there is a provision, already quoted, according to which this exchange is indeed expected to take place. Section 6 :

> The Bureau of the Mint shall receive, subject to a "brassage charge" of 1 per cent. . . . all gold bullion offered to it, and shall pay for the same with "gold bullion dollar certificates" described hereinafter . . .

And Section 7 inverts the case :

> That the Bureau of the Mint shall receive all gold bullion dollar certificates tendered to it and shall forthwith pay for the same, *dollar for dollar*, in standard gold bars at the rate of $1 for the number of grains of standard gold in the gold dollar of the United States . . . and in lawful money.

Obviously the gold thus bought and sold would go to and come from the reserve, altering its contents and thus disturbing its ratio to the outstanding paper money. A very important point about this provision is that the exchange proceeds from the public and not from the managing authorities. It remains to be seen whether the public will be able to act up to the expectations.

We have to make out, in the first place, what can induce the public, in which I include the banks, to offer gold to the Bureau of the Mint, or to demand gold against certificates. Gold is offered to the Mint when the holders of gold expect to profit by the exchange, that is when gold is depreciating in terms of paper money; in other words, when gold loses its usefulness and people would rather be rid of it. How can this come to pass? What is this depreciating gold, money or a ware? It is demonetized; but that makes little difference, as pointed out above in more than one connection: gold is always money. Also the Bill speaks of the "purchasing power of gold," by which gold is invested with the money character. Gold, then, is money. But why should gold be offered for money if it *is* money, neither better nor worse than the other kind? And by the Goldsborough plan the gold dollar is to be kept at a parity with the goods dollar: when gold depreciates the gold dollar is enlarged, so that there can be no advantage in handing gold into the Mint; the holder has to bear the loss of depreciation whether he keep the gold or surrender it. Thus we are led round to the conclusion again that the reserve can play no part at all, and would seem to be perfectly useless.

It must be intended to render some kind of service. What better service could we demand of it than to be the refuge of depreciating gold and the conqueror of appreciating gold? This demand cannot be met. Gold is money, whatever the shape in which it presents itself. It depreciates when it is not allowed to act as money, and as money it depreciates when money generally depreciates. Depreciation of money is another term for the rising of the general level of prices, and it follows that gold should flow back to the reserve in a period of rising prices. However, that cannot be under the assumption that it is exchanged, and that at par, for paper money; neither would such an exchange produce any effect on the price movement. But gold is not only money, it is also a ware. When it depreciates in its capacity as money, it does not do so in its capacity as a ware; as a ware it rather appreciates, wherefore the holder of gold would not think of handing it back to the reserve against depreciating paper. He does not need any refuge for his gold, because gold cannot depreciate.

For curiosity's sake I will point out that by the authentic interpretation the rise of the Index-number is countered by the increase of the weight of the gold dollar. That is to say that gold is supposed to be depreciated by the going up of the price level—gold sharing the fate of lawful money, *as being money pure and simple, and nought but money.* Here we touch upon the central misconception from which the gold dollar fallacy flows.[1] According to the interpretation yellowbacks, that is bank-notes or lawful money, would have to be withdrawn from circulation when the gold dollar is increased. This would coincide with the situation in which, according to the theory of the scheme, gold would flow back to the reserve in exchange for paper money, and thus the number of yellowbacks would be increased rather than diminished.

However, it is impossible that the scheme should work. Gold will not be delivered to the reserve, nor will it ever be withdrawn from the reserve, on the assumptions of the scheme. These assumptions, we are to understand, are that whether the reserve convert paper money into gold, or gold into paper money, it shall make a profit out of the

[1] The error is derived from the traditional gold standard theory, by which gold was only money. See below, p. 242.

person demanding conversion. It is expected that the public will deliver gold when gold has been cheapened: give a larger lump for a goods dollar. The profit would appear when the tide has turned and gold has appreciated. In this juncture it would seem that the public ought to try and get gold out of the reserve, which they can by tendering gold bullion dollar certificates. The gold dollars which the reserve would sell under these circumstances would be smaller lumps of gold than those which it received when gold was handed in, and the difference would make up its profit. It needs no proof that the public will not respond to the tune of this piper. The gold dollar scheme resembles Gesell's proposal regarding the sale and purchase of securities: the mere fact that the provision concerning the exchange of gold, if carried out, would invariably result in a profit to the Treasury renders its execution impossible.

Or suppose the process to be inverted: gold is returned to the reserve when it is dear—at its dearest, just before it begins to decline—and called out when it is at its cheapest. In this case the reserve would receive small lumps of gold and part with large lumps for its goods dollars, and so always sell at a loss. Curiously enough this is no more possible than the first assumption, simply because nobody would think of giving dear gold to the State. Besides, if gold were actually returned to the reserve under these circumstances its price would rise further—seeing that the article would become scarcer—and the seller of gold would make a losing bargain.

We have floundered about in the experiment without ever touching the bottom. It has no foundation. The basis of currency must not be looked for in gold and gold reserves, but in something very different. Gold enters into the superstructure of the currency, both as a ware and as money. But the chief reason why our endeavours to comprehend the mechanism of the gold dollar have been foiled must be traced to the fact that the price of gold, in a stable currency, cannot vary. It is impossible to create a gold dollar of varying weight and constant purchasing power. The two terms exclude each other. The price of gold varies while the general level of prices, in terms of money, varies; for gold is money, the incarnation of money.

Another lesson to be drawn from our examination of the Bill concerns the interference of the Treasury (an institution of the State and an instrument of the Government, of party politics). Gold and gold certificates, we have shown, cannot be exchanged under the regulations of the scheme. The author must have been aware of this, otherwise he need not have devised any action by the Treasury. As the case stands it is the Treasury that is supposed to issue and call in the bank-notes. But we ask—assuming that it could do the thing—in return for what does it give and receive them? Bank-notes are means of exchange and ought to be given and taken in exchange for something else—and for something real at that. There is no other fair and decent procedure. We have had a good deal to say about the proposals of the reformers to manipulate the National Debt and the taxes; Mr. Goldsborough, for his part, has nothing to propose whatsoever. For surely the provision of Section 10, to the effect that Government deposits in national banks shall be withdrawn or added, leaves us in the dark. By carrying it out the State would merely shift its assets from one place to another. If not so the deposits which are withdrawn would have to be destroyed, while those sums which are added would have to be manufactured out of nothing, which is not encouraging.

We have been baffled in our attempts to get at the manner of the variations of the gold reserve. As a matter of fact, under a stable standard of currency and short of great natural catastrophes, the reserve may be expected to remain an almost constant quantity. Gold would be handed in if it depreciated, and taken out if it appreciated—and that is exactly what used to happen under the unstable traditional gold standard. In a stable currency gold cannot be detached from the parity. Nothing can happen to induce anyone to try and convert either paper into gold or gold into paper. The owner of gold can in no circumstances realize a greater profit by selling gold than by keeping whatever gold may happen to have got into his possession. If gold depreciates as money it appreciates as a ware, and all that the owner of gold, to avoid a loss, has to do, and can do, is to reserve his gold coins. But in doing so he helps to forestall the appreciation of gold, so that his speculation bears no fruit. (Gold is replaced by paper money and so rendered dispensable.) However, no sort of speculation with gold can yield a profit, if hoarding it does not. Selling gold for bank-notes gives to the seller what can neither appreciate nor depreciate in terms of gold ; so what inducement could there be for such a transaction ? The same applies to the purchase of gold for bank-notes. Thus the speculator with gold has only one chance left : to reserve his gold when it would depreciate (as money) if he did not reserve it as a ware, and to circulate it when it would depreciate (as a ware) if he did not circulate it as money. If it is objected that he who hoards gold forgoes the interest on the sum in question, I will say that nothing prevents him from depositing the amount at his bank, which can safely engage to repay the deposit in gold, if such clause should be stipulated for. Consider the circumstances in which gold can be reserved : there is an incipient excess of cash in circulation, which tends to force up the rate of interest. The gold deposited at the banks will assist in depressing the rate of interest : instead of drawing bills for discount the business man borrows cheap gold at his bank, and thus the gold deposits flow into the circulation almost immediately, displacing the bank-notes which otherwise might have been issued against bills. The automatic play of free, untrammelled gold makes speculation with gold look like a silly pastime.

This way of considering the mechanism may contribute to elucidate the reason why under a fluctuating standard of currency the gold reserves have alternately grown and diminished. By the imperfect traditional gold standard the parity between gold and paper money was theoretically fixed, but practically it was not assured and far from being maintained. The fact that gold was ever withdrawn from the reserve is a direct proof that gold was better money than paper ; gold would buy more goods than bank-notes. And vice versa, when gold accumulated in the reserve it must have been because paper money was better than gold. In order to understand why it is that under the stable interest standard such departures from the parity cannot happen, it is necessary to trace the influence of interest on the movements of gold and paper money respectively.

Gold was induced to leave the reserve when the rate of interest was high—at the close of a period of rising prices ; it returned to the reserve when the rate of interest was low, at the close of a period of falling prices. In trying to discover the reasons of this we have to remember the uses of gold. In prosperous times, when business expanded, the prosperous debtor class reached out for gold in the shape of articles of luxury ; the increasing demand for gold raised its value, as a ware. But while its value was enhanced, its price continued unaltered, most unnatural of phenomena. For gold could not go up in price because its price was legally fixed, and the result was that it was cheapened in terms of other values. By being tied to its fixed price—crucified, so to speak—gold was rendered altogether defenceless ; it was unable to appreciate as a ware when it was depreciated as money. It soon became so cheap that the gold-mines were forced to reduce their output. However, a point was reached when gold gave out, and could not be procured at a parity any longer in the open market. Then it was that the reserve began to be drained for export. It was because the supply of gold was running short in consequence of increasing consumption and decreasing production.

In what respect could this development be attributed to interest; what had interest "to do with the case"? It was the raising of the rate of interest that made prices go up, thus depreciating money and, along with it, gold. Had the rate of discount been kept fixed and low the easy supply of cash would have prevented the creation of new bank-notes through the discounting of bills of exchange. Events would have taken a different turn: no inflation of prices, no cheapening of gold, no increase of its consumption and simultaneous decrease of its output, no need of any gold exports. When gold is not tied to a price it is not defenceless, and this being so it will successfully oppose attempts to cheapen it. (Why the level of prices cannot rise when the rate of discount is fixed is fully demonstrated above, p. 195.)

Gold, then, was withdrawn from the reserve when it broke away from the parity to appreciate in terms of paper money (thus reasserting its character as a ware), and it did so in the teeth of the law which fixed its price. But it was unable to do so until things had proceeded to a point at which the situation became unmanageable and necessitated a reaction. The only way to prevent such an eventuality is to give gold a chance to react as soon as ever it is menaced by depreciation.

Now let us also outline the opposite course: gold accumulating in the reserve. It can only come to pass when people are forced to be rid of their gold and procure money, i.e. when money is more useful to them than gold. This is the case when they have certain liabilities which they can only meet through the sale of dispensable—and saleable—goods. In a period of depression the only saleable article—how many soever may be dispensable—is gold, and it is in this predicament that those who bought gold trinkets during the boom are compelled to sell this gold again. After a certain reduction of commodity prices the output of gold paid once more, and more gold was mined. Gold flowed into the countries with a falling level of prices ; exporters were paid in gold, which they gave to the Bank in redeeming their bills of exchange.

Depression is not likely to happen except as a reaction against a boom that has preceded. Let us, however, assume that gold began to be short without the provocations indicated above. Under the traditional gold standard such shortage was bound to produce a shortage of money, and consequently deflation. It was this movement that used to force debtors to part with their gold ornaments (also creditors were often forced to do so when their debtors failed and their fortunes were ruined). But prices cannot decline if the rate of interest is maintained, as it is by my scheme of the interest standard ; they are upheld by additions of paper money procured at the Bank of Issue against discounts. The shortage of gold does not result in a shortage of money, because a more liberal use of the substitute of gold, paper money, makes up for the deficiency. Yet at the same time the production of gold will be encouraged, so that the shortage will find its real and natural remedy.

I have been at some pains to account for what used to happen so frequently. Apart from logical reasons we are taught by empirical facts that gold used to be drained out of the reserve whenever the rate of interest was abnormally high, and in no other circumstances. A very good proof of this is to be found in the rule according to which the English Bank Act cannot be suspended without a 10 per cent. bank-rate. A suspension of the Bank Act is only conceivable as a means to protect the gold reserve, and so it is clear that a high rate of discount is the condition in which a gold drain is likely to occur. Now consider the case of an experimenter in a laboratory who has discovered that the stopper will invariably fly out of his bottle when a certain temperature is passed; he will not let the temperature go beyond the point if he does not want the stopper to fly out. Quite otherwise is the logic of the guardians of the gold reserve: knowing by experience that gold only tends to escape when the rate of discount is above, say, 5 per cent., they will raise the rate in order to protect their treasure. Surely, the safer and more logical proceeding would be to keep the bank-rate below the danger line. From all these observations it also follows that a fixed rate of discount must make the flow of gold to and from the reserve impossible, simply because no one could conceivably gain anything by offering gold to the reserve or demanding it of the reserve.

It may be asked whether gold reserves will not grow when more gold is being produced, thanks to new discoveries or better facilities. No gold can ever be produced as cheaply as the gold of the reserve, and therefore no gold will be produced that could disturb the peace of the gold reserves. That is to say that new gold will be mined only to the extent that it pays to do so, on the basis of the invariable price of gold. The gold production will adapt itself to the needs of the market. If it supplies more than is demanded, the price of the article will suffer, and gold will be forced to serve more generally as a circulating medium—which amounts to saying that it will not go to augment the reserve, being kept busy. If, contrariwise, the output of new gold is insufficient, the price of the metal will rise, and gold will be relieved, to that extent, of its service as a circulating medium—which again amounts to saying that no gold need be withdrawn from the reserve to keep the market supplied and the price of the metal steady.

The long and the short of the case is this: the reserve need not be approached, all the changes that take place in the monetary organism being compensated for by the gold outside the reserve—in which the undug gold is included. Hence the reserve will continue as invariable as the price of gold. Of course this state of things will not produce itself on the very day when the new currency law is enforced. The gold reserves must needs be fairly equally distributed among the trading nations before gold can keep the peace. And further it seems to me that gold will always follow the flow of population, so that the national reserve would grow and lessen along with the increase and diminution of the population—or else the other way about: if gold cannot follow men, men must follow gold, a country which is depleted of gold also losing inhabitants, a country gathering gold also acquiring inhabitants.

If this surmise is founded in fact and sound logic, it would follow that the gold reserve, both national and private, of a country is in a certain natural proportion to the population, to its humanity. Here is a problem for the metaphysician to speculate upon. Does sleepy Spain contain as much gold per head as active England, and backward Russia as much as forward Germany? Is gold an organ of humanity, like the blood that flows in the veins of individuals?

One word more as to the uselessness of the everlastingly buried gold in the cellars of the central banks. These reserves are no more useless than the foundations of a house, which are also buried in the ground. The gold in circulation would be unable to perform its function as a stabilizer if its volume were increased by the weight of the reserve gold. Business could not absorb the whole, and the excess would be a menace, a dead weight. The most useful things are, perhaps, not those that fly about as busy-bodies; "they also serve who only stand and wait."

This subject will be touched upon once more at the end of the next following section.

3. The Creation and Destruction of Credit by the Bank

In describing the "pre-war monetary system," Mr. Withers represents the Bank of England as being the creator and destroyer of credit. He does not stop to examine and show why at one time the Bank would expand, and at other times contract credit. He implies that it acted upon the initiative and discretion of its own managers, and was the sovereign ruler of the money machine. He credits the Bank of England with having created, out of the Bank Charter Act of 1844, an almost perfect system of a managed currency. Here is a passage to illustrate what he considers to be the superior excellence of the system (p. 107):

> . . . especially when we remember that by the altogether uncontrolled right of the Bank of England to create credit in its books for any borrowers to whom it chose to lend, and the convention which allows the other banks to regard a credit at the Bank of England as "cash."

I had thought that "Britons never, never, never would be slaves." The "altogether uncontrolled right" of the Bank of England, an entirely private concern, would have been an intolerable tyranny if it had really existed and been exercised. Indeed, these rulers of the banking machinery have been erected into a sort of Olympus by scare-mongering reformers, who have represented them as holding governments in the hollow of their hands, and exercising discretionary sway over the fortunes of the business world. No doubt the worthy gentlemen in charge of the machine have been not a little flattered and exalted by this interpretation of their position. Who would not be proud of being charged with having mismanaged the business of a country? It implies that he was entrusted with its management, that he has played a great part. But was not the Bank the managed rather than the managing party? The close reader of Mr. Withers will discover many a passage contradicting the legend, which he so assiduously helps to fix in the belief of the public, that the Bank of England did the real managing. Indeed, in his endeavour to exonerate the Bank from the charge of having inaugurated the crisis of 1920 by a deflationist policy, he goes so far as to show that the Bank's policy was nothing but the enforced consequence of a purchasers' strike, which began with the boycott of Japanese goods by the Chinese patriots. Thus it appears that the high and mighty institution yclept Ye Bank of England is prompted in its measures by the whims of pigtailed Chinamen.

It may be worth while to examine an instance or two of the managing with which Mr. Withers has credited the Bank. Take the case of the contraction of credit, as stated on p. 34 :

> On the other hand, when the Bank of England reduces its securities, either by selling stock or by inducing its borrowers to liquidate their loans by charging them a higher price for them, which it does by raising its official rate of discount, the effect is to contract the basis of credit and to cause a still greater contraction in credit, so that the proportion of cash and credits may be maintained more or less on the usual level.

This is altogether after the manner of Mr. Keynes's "managed currency" : a Bank of Issue acting consciously and deliberately to bring about a calculated result. We find the same terminology : "the basis of credit," meaning cash in its various forms, and credit proper, as expressed by the rate of discount. Mr. Withers, critic of reform schemes, has imbibed, without realizing to what extent, the conceptions of the reform schemes, and the best part of his criticism is directed towards proving that what the new schemes aim at was already secured, and more efficiently, by the old practice. Now my contention has persistently been that currencies cannot be managed. I shall make use of Mr. Withers's proofs in trying to show that the belief in the Bank's capacity to direct the course of events is a delusion and the source of very serious practical mistakes.

The passage quoted implies that the Bank sold stock with the purpose of creating a certain desired situation (desired by certain interests, that is to say). It does not stop to inquire about the inducement on which the Bank acted, nor about the conditions which were necessary to render the particular action possible. Like Mr. Keynes and all the reformers, Mr. Withers assumes that the Bank is able to sell when it pleases and on its own terms. Of course the Bank would make good use of such a fine opportunity, never selling except at a handsome profit. But the Bank is no more favoured than humbler bargainers ; it is forced to abide its time and sell when it can and must. It can only sell to a market willing to buy, i.e. to people speculating on a profit to be made out of the exchange of cash for securities. The Bank will find it impossible to sell stock while the market is rising ; if the impossibility has not been proved by events, it can only be because the Bank has never yet made the attempt. The Bank is forced to wait for the turn of the tide. Not till credit is already contracted is the Bank able to contract the basis of credit. When the holders of cash have lost their belief in the continuance of the boom, i.e. when they have ceased to credit the chances of real property, they will try to exchange their cash for Government stock, for securities. The demand for stock must be afoot before the Bank can take any measures ; it is not the Bank which induces the public, but rather its moves are induced by what happens in the market. The Bank has to follow suit in a game not of its own making.

How little the Bank was able to manage, to control the situation, is most aptly demonstrated from the manner of its action when gold flowed from its reserve. The case is neatly presented by Mr. Withers. Gold was given out in exchange for bank-notes, " which were forthwith cancelled, because their backing had gone. The consequent reduction of the Bank of England's reserve, and of its proportion to liabilities, would immediately be noted by the market as likely to cause the Bank, if the process went farther than it thought prudent, to raise, by an advance in bank-rate, the price at which it would grant credit." The reason for raising the bank-rate, then, was to stop the drain of the Bank's reserve, the managers posing as the dragons posted in front of the cave where the hoard lies buried. But was such action necessary ? Was it really the belching of dragon fumes which produced the check on the flow of gold ? What inference could the market draw from the fact that gold was being demanded of the reserve and shipped abroad but that the supply of money was running short ? Mr. Withers himself very clearly expresses this inference where he writes (p. 171) :

> In old days when the bank-rate went up it meant that a gold drain was occurring or expected, and a gold drain meant that there was less money in the country, and that borrowers, whatever price they liked to pay, would have to have less.

So it was not necessary to warn borrowers by raising the rate ; the observation that gold was escaping was sufficient warning. Moreover, the statement of the case is incomplete. When the bank-rate went up it did not always mean that a gold drain was occurring, nor even expected. In the nineties the rate of the Bank of England was down to 2 per cent. It must have been raised to its normal level in many successive stages, when the earlier moves surely could not signify that the gold reserve was in need of protection. In the second place the occurring of the gold drain was signalled by the periodical publication of the Bank's returns. Nay, more : the business world must have felt the pressure of circumstances even before any figures were published, since they were being induced to do the unusual thing, i.e. to obtain gold at the Bank. We may be quite certain that the market did not need the clumsy hint from the Bank to be made aware of what was imminent. The market would have responded to the situation in exactly the same manner if it had not been made to anticipate a rise of the bank-rate. That the rise itself was a mistake has been pointed out above (p. 241).

For what consequences could it expect from the running short of gold but that prices would soon begin to decline? How could the market react upon this intimation except by slowing down its activity? By withholding or reducing orders, by realizing outstanding debts, by contracting credit? Fewer loans would be applied for, sellers would endeavour to be paid in cash. If the Bank, under the circumstances, had acted logically it would have rather reduced than raised its rate. Its action only served to scare the market and aggravate the difficulty. That, moreover, it failed to produce the desired effect is shown by the same champion of the gold standard, Mr. Withers, where he explains why borrowers were forced to borrow the Bank's notes in spite of an exorbitant rate (p. 158):

> Later on, when the wind blew on the other cheek and prices were falling like a stone . . . the price of money was equally irrelevant for a different reason. Producers and merchants had to have it then, because they could not sell their goods . . . and so those who had goods on hand had to have money whatever it cost, because they simply could not sell them, and were forced to carry them on credit. And so the Bank Rate and the Treasury Minute, which was supposed to set a limit to the note issue, had no visible effect according to the testimony of available figures in contracting credit, because credit could not be contracted until the stock of unsaleable goods was gradually worked off.

Mr. Withers argues that the measure failed to take effect because the gold standard was not then operative; but obviously the reasons given must be valid in all circumstances: when prices fall "those who have goods on hand have to have money whatever the cost." Of course the high rate of discount cannot but be an extra calamity for them; its raising simply adds insult to injury.

In more than one connection Mr. Withers rightly points out that in a period of shifting price levels the moving of bank-rate is impotent to check the tendency of a rising or a falling market. He says (p. 169):

> Surely if a man thinks that by buying a security, or a commodity, or a currency, he gets a good chance of being able to sell in a few months at double its present value, he will be quite ready to pay 20 per cent., or even 50 per cent., for any money that he wants to finance his gamble.

And p. 253:

> But when the wind is on the other cheek, when depression prevails, and it is desired by lower money rates to induce enterprise to tuck up its sleeves and get to work, no such success can be expected with any confidence. If people think that any business that they do is likely to involve them in a loss, they would evidently be born fools to do it, even if the money needed was lent them for nothing.

Mr. Withers quotes a number of authorities, English and American, in support of the view which he himself endorses, namely, that "in normal times a few turns of the bank-rate screw might certainly be counted on sooner or later to produce the desired effect." "Sooner or later," and "a few turns of the screw": why, of course. But it is quite certain that the first and second and third turns will produce the contrary effect. When at the end of a depression bank-rate is raised again, prices will move upward. Just look at the records of such countries as England and Switzerland within the last few years: the raising of the rates of discount was in due course followed by an advance of the price-index—all in spite of the fact that in both countries a special effort was made, officially, to bring prices low (back to the dollar parity), without which prices would have risen even more vigorously. The record of the United States currency in the year 1924 proves the case from the opposite side:

prices declined in the wake, and in consequence of, the reduction of the discount rate. Mr. Withers himself furnishes a proof of the same, where he informs us that (p. 161):

> Banking figures went on expanding long after the rise in bank-rate, and only contracted appreciably when bank-rate was being reduced again.

I could not wish for a better corroboration of my contention that the traditional theory and practice of discount puts the cart before the horse. The raising of the rate stimulates prices to advance, the lowering of the rate stimulates prices to decline—up to, or down to, the point when the machinery is either so overstrained, or so exhausted, that it collapses and ceases to respond to pressure. I cannot think of an apter illustration than that of the steam engine: while the pressure is being increased its speed grows and grows until at last something breaks down and the race suddenly comes to a standstill; vice versa, while the tension is being reduced the motion must slow down more and more till finally new forces come into play. The rate of interest indicates the degree of the tension in the economic machine, and interest itself is the motive force.

To those who feel inclined to turn the point of this statement against my own solution I would object that fixing the rate of discount is not the same as fixing the rate of interest. The latter has sufficient scope left to it for exercising its function as a regulator and index. Narrow as the limits of its fluctuations may become, its tendencies and actual movements will mean, and be heeded, all the more.

It is grossly to exaggerate the importance of conscious and would-be logical measures to think and say that, short of them, the course of events would take a different turn. The movements of gold, which bank-rate policy is supposed to control, have a significance far beyond the reach of such policy. They occur because something else has occurred.

When gold has flowed into a country for a period, in consequence of under-consumption, something must happen to drive the surplus gold out again; and vice versa, after gold has been drained out of a country, owing to over-consumption, the moment is bound to come when this cannot endure any longer and a course in the opposite direction is begun. Bank-rate policy is no more than La Fontaine's *mouche du coche*: the fly which fancied that its buzzings were the action that set the coach going.

Rightly considered bank-rate policy is nothing but a repudiation of the fundamental principle of the gold standard. Mr. Withers quotes from and comments upon the *Cunliffe Report* to this effect (p. 96):

> Under an effective gold standard all export demands for gold must be freely met. This observation once more shows that the Committee's whole basis was free export of gold. It went on to point out that some machinery must exist to check foreign drains when they threaten to deplete the gold reserves; that the recognized machinery for this purpose is the Bank of England's discount rate; and when the exchanges are adverse, and gold is being drawn away, it is essential that the rate of discount should be raised relatively to the rates ruling in other countries.

Note well the stresses of the first sentence: all demands—freely met. Then note how everything that follows insists that all demands must not be met. To apply a system which is based on such gross and flagrant, though apparently unsuspected, inconsistencies is downright dangerous. I do not object to a country trying to protect its gold reserves by measures which cannot but be directed against the interests of other countries; in matters of gold it is foolish to preach altruism; but I object to the deception that gold is allowed to move freely when the whole machinery of the Bank is devised to impede its freedom.

The practitioners of the gold standard, it seems, do not understand why gold is drained out of a country. Gold quits when it is not usefully employed, and it goes to where people have some use for it.[1] Nothing could be more natural or more wholesome, and nothing could be more unwise and injurious, than to interfere with the movements of gold. I have stressed the point again and again that gold deserts from countries in consequence of over-consumption or under-production: gold is no longer buoyed up sufficiently and would be left lying high and dry if it did not take flight. If it is allowed to go freely and to the full extent its drain will work the natural cure. Imports have exceeded exports, and the shipment of gold abroad has to make up for the deficiency and restore the balance. As the gold coins disappear from circulation, people are warned that their wealth is disappearing. If the process continues, it becomes more and more difficult to procure gold, because gold coins will be withheld by the public. This withholding signifies that the public buys less and consumes less: over-consumption is thereby checked. But that is not enough; we want to produce more. The disappearance of gold forces us to do so; an empty purse is a powerful stimulus to production. When gold leaves the country all the productive forces are called into activity to draw the fugitive back again. But the interference of the Bank, according to the procedure sketched by the *Cunliffe Report*, cannot but destroy the chances of business to succeed in recovering its productivity. After gold has escaped and left the circulation of money weakened, the raising of bank-rate with the purpose of scaring business into surrendering paper money was and will be the straight road to the killing of enterprise. Indeed, it is only owing to the fact that the measure failed to produce the desired effect that the damage wrought was not more serious.

[1] "When, therefore, the state of prices is such that the equation of international demand cannot establish itself, the country requiring more imports than can be paid for by the exports, it is a sign that the country has more of the precious metals or their substitutes in circulation than can permanently circulate, and must necessarily part with some of them before the balance can be restored."—Mill, *Principles of Political Economy*, Book III, chap. xxi., § 1.

But the alarums and excursions of the guardians of Pluto's treasure were a false alarm. Nothing could have happened to the gold reserve. Nothing, most assuredly, can happen to it under the regulations of the discount standard. I must try to convince my reader so fully as to allay his instinctive apprehensions, and I shall not, therefore, scruple to demonstrate once more the mechanism of the system. Let us assume that a heavy gold drain has already taken place, so that gold coins have ceased to circulate. This assumption is necessary if we are to admit the necessity for business to apply for gold to the national reserve. Gold is delivered in exchange for paper money, and at par—for we are not going to impede by any restrictions whatever the movements of our gold. The result is that the volume of paper money begins to shrink. How then are the heavy imports of goods (the fruit of the gold exports) to be absorbed by the public, if there is a shortage of the means of exchange ? Bills of exchange are called to the assistance, you say ? But how could that be, since goods are paid for, abroad, in gold ? The money received against the bill has to be returned immediately against gold, and therefore no increase of the note issue can be brought about. The consequence is that the money shortage effectually compels the public to defer purchases and to retrench, which will also compel merchants and manufacturers to bring their stocks to market and reduce their prices. Here we have collected all the data needed to arrest the depreciation of money (gold) which has been the sole cause of the supposed flight of gold. It cannot take place because it is against nature that the gold drain should be carried to the point where it would seem necessary to approach the gold reserve.

As I tried to demonstrate in the previous section the gold reserve under a stable standard of currency is likely to be maintained practically constant. However it is necessary to face the possibility of gold having to be withdrawn for export. Facing it we shall understand all the more clearly the absurdity of any clauses to restrict the export of gold out of the reserve.

What are the real meaning and uses of the thing? A reserve is wealth laid by against an hour of need. Suppose an earthquake or some other general catastrophe to have devastated a country: if the people are to be preserved from starvation large imports are required at a time when exports must needs fall to zero. Shall the Bank, to which the nation's gold reserve is entrusted, be allowed to stop the drain of gold, to scare off merchants who would like to borrow the means for meeting their foreign acceptances? Surely the clause, in such an emergency, would be suspended without much scruple. Then why make it a cornerstone of the Bank's constitution? In normal times it is unnecessary, and in a time of stress it is immediately overthrown.

There is another point to be considered. It is a proven fact that the crisis of 1907, which caused so much embarrassment to the Bank of England and such enormous losses to British business, was a machination of a few American banking magnates. It was the clause here under discussion which made it impossible for the Bank of England to meet the needs of its clients in this emergency. Without its restrictions nothing would have been easier than to supply sufficient paper money to make up for the gold withdrawn from circulation by rapacious speculators, and so to save the structure of prices and tide business over the shoals. Of course the speculation of the sharks would not have succeeded quite so well as it did. But many an honest merchant's integrity would have been saved, and unemployment would not have oppressed the working classes.

If in one place I have made out a case for the assumption that the gold reserve will keep at an approximately constant figure, while in another connection I have shown how national gold reserves are formed and dissolved, I have not made myself guilty of any inconsistency. A growing organism gains weight, a declining one loses weight. A full-grown healthy person who lives wisely maintains a pretty nearly constant weight, shunning fat. Should he contract an illness, from which no one can be immune, he may lose 10 or 20 or 30 pounds ; a few weeks after recovery will suffice to restore the normal weight. This analogy holds good for the gold reserve, which is like a living organism. It must not keep changing its weight if it is to be uniformly efficient and healthy ; but also it must be allowed to give off weight in a period of distress. There can be no danger in letting go as much as will and can go, and it will very speedily regain its pristine portliness when the organism is able to resume its normal functions.

4. The "Self-Liquidating" Bill of Exchange

My system is based on the principle that money is to be issued *ad libitum* against commercial bills of exchange, and that no considerations whatever shall limit these issues. I have already dealt with certain objections to this procedure, and here return to the subject in taking up the case as presented by Mr. Withers, pp. 89–92. He discusses the proposals of a reform of the Bank of England made by Sir Edward Holden, a leading banker. I quote in full :

> The principles for which he contended are those on which other national banks of issue, according to his account of them, worked. These principles are :
>
> (1) One bank of issue not divided into departments.
>
> (2) Notes are created and issued on the security of bills of exchange and on the cash balance, so that a relation is established between the notes issued and the discounts.
>
> (3) The notes are controlled by a fixed ratio of gold to notes, or of the cash balance to notes.
>
> (4) This fixed ratio may be lowered on payment of a tax.
>
> (5) The notes should not exceed three times the gold or the cash balance.

It will be seen that on these principles indefinite expansion can be secured. The basis of your note issue is not to be as it was in our case before the war—gold except for a fiduciary issue, the amount of which was fixed—but bills of exchange with a certain proportion of gold, which proportion again can be reduced apparently at will by the payment of a tax. Thus, if the community will only draw enough bills of exchange and pay a certain amount of tax, it can have as much legal tender currency as it thinks it wants. Much was said, in the discussion that Sir Edward's proposal aroused, about the beauties of the bill of exchange as a "self-liquidating" instrument; and it is true that a bill drawn on a solvent acceptor against goods on the way to market has outstanding merits as a short investment. But it was not quite right to assume, as happened, that currency based on bills of exchange cannot be multiplied to excess because there must necessarily be goods behind every bill, and so currency could not increase faster than goods. In fact bills are often drawn without goods behind them, on the credit of the parties; and even if it were possible to exclude such kites from being used as a basis of currency, it is still possible that several bills might be drawn against one parcel of goods. Goods generally change hands many times on their way from the producer to the final consumer, and if they are paid for each time that they are sold by the drawing of a bill at three months, and all these bills could be discounted at the Bank and new notes created against them, a shipload of cotton or copper might easily have ten times its value in notes outstanding on its responsibility. And sometimes when goods do not find a final market bills drawn against them are renewed; and when this does not happen there could always be fresh batches coming forward produced in the same way.

Under these arrangements, if carried to their logical conclusion, the task of those who have to maintain the country's gold reserve, on the assumption that a gold reserve is a necessary basis of a monetary system, would be extremely difficult. In one of its exuberant moods the commercial community could demand the creation of fresh currency as fast as it liked, continually forcing up prices, and so putting further nominal profits into its pocket as the process continued. . . . The end, when the system is worked to its logical conclusion, has been shown in Russia, and other countries where the currency has become practically worthless.

A question by the way: were the notes issued against bills of exchange, in Russia and those unhappy other countries? No, they were not, and the "logical conclusion" to which Mr. Withers has carried his argument therefore lacks logic.

In what particulars does the Holden scheme differ from mine? The former provides for a "fixed ratio of gold to notes," which, altogether after the fashion of the real gold standard style, may be unfixed when circumstances seem to call for a special indulgence. My scheme does not fix the ratio of gold to notes. It fixes the rate of discount—and fixes it really and truly, in good earnest, once and for all, least to be moved when circumstances most seem to demand its being adapted. Now let us examine whether "on these principles indefinite expansion can be secured." The main point is to discover whether or no the scheme insures the self-liquidation of the bills of exchange. For obviously, if money is only issued against bills, and if these bills do liquidate themselves, it must be impossible to force out an excess of currency and bring about inflation.

The first question to be answered: is there any chance of too many bills being presented for discount? It will be readily admitted, I think, that in my scheme no influence can be exercised by anything that may be happening to gold. When the connection between gold and the currency has been severed, we have become independent of gold.

An increasing efflux of gold will not stimulate, at least not over-stimulate, business, nor will the outflow of gold produce a chill and a scare, as happened under the old gold standard, and would happen under the Holden plan (which indeed is a copy of the defunct German gold standard law). When gold comes in greater force, everybody will know that it is going to displace bank-notes, instead of calling forth new bank-notes, as was the case hitherto. Gold, in these circumstances, tends to depreciate through over-supply; it is forced to serve as money, and in doing so it narrows the field of paper money, which cannot be cheapened thanks to the law which fixes the rate of discount. Moreover, we ought to remember that gold flows in when the country is threatened with a depression and prices are tending downwards. So an increase of the volume of circulating money can only produce a beneficent effect: stimulate consumption, and cause a prompter turnover of the goods produced. Depreciation of gold does not begin until the output of gold becomes excessive; but when this situation does occur, the reaction on the part of paper money will prevent inflation in the manner described above.

When gold is found to be leaving the country, everybody will know that, far from forcing the Bank to cancel notes, as by the rules of the old gold standard, this movement can be freely compensated for, if necessary, by new issues of bank-notes: gold tends to appreciate, and to resume the character of a ware, by which the chances of paper money are improved—the business world applies to the Bank for notes against bills of exchange. Here again I must remind the reader that gold will begin to be exported when consumption has outstripped production, so that a reduction of the volume of cash is needed to prevent prices from rising. Thus the efflux of gold money is a natural reaction tending to diminish the circulation of money and to forestall an impending or to check an incipient rise of prices. Even under the old gold standard the drain of gold need not have created a shortage of money, with the consequent fall of prices; but the money shortage was created by the withdrawal of notes, which cumulated the effect and destroyed the outlook of the business world. Under the interest standard notes will not be withdrawn, credit will not be paralysed, and the departure of gold will be a wholesome working off of an excess of gold, and an automatic restoration of the equilibrium.

This automatism must also extend to other regions and oppose, stroke by stroke, every tendency towards either inflation or deflation. When borrowers are too aggressive, and the rate of interest inclines to advance, creditors will automatically react in self-defence. Under the false gold standard they were defenceless, and their only way of escape was to desert to the enemy, to throw themselves into the chase, to try and get hold of money in order to buy. Under my scheme, the true and natural gold standard, nothing of the kind. When the economic barometer, which is the price of securities at a fixed rate of interest, indicates an inflationist atmosphere, the holders of securities will not try to sell, to convert bonds into shares, but will stick to their guns, knowing that the fixed rate of discount will be a sufficient protection. The rate of interest cannot be forced more than a trifle beyond the official rate, and so a depreciation of securities is precluded. By this same fact, now, an appreciation of commercial stocks and shares is likewise excluded. Hitherto the gain of the latter was squeezed out of the substance of the former ; now when bonds cannot be squeezed shares cannot be fattened, or inflated, with their juice. When you fix one end of a see-saw, the other end is equally well fixed, and when the price of bonds cannot fall, the price of shares cannot rise.

Everything considered there is no fear lest too many bills should be discounted in the normal course of affairs. But Mr. Withers mentions the possibility of certain abuses : bills may be drawn without goods behind them. This was practised under the old system, yet I have not seen it mentioned as ever having been the source of inflation. It is a trick to which impecunious debtors will resort, and as this species is not likely to become extinct, I shall not deny the possibility of the abuse being continued. Its effect, however, cannot be less negligible than hitherto. Next there is the possibility that one and the same parcel of goods, in passing through the hands of a number of middle-men, may be drawn against not once but many times, which would result in an over-issue of notes. This abuse will not be practised to a dangerous extent except in periods of inflation, when goods are being withheld with a speculative purpose, because they cannot be replaced by fresh output. The more a currency becomes stabilized the less will the abuse pay, and the sooner will it be discontinued. Besides, there is this to be said : bills of exchange have to be redeemed (sooner or later even those which have been renewed), and if many have been drawn against one parcel of goods, just so many will have to be paid in the end, so that all the money issued against them will return to the Bank. Inflation is thereby averted.

Generally speaking the critics of the schemes for stabilization are apt to apply war-time experiences to the examination of a mechanism which in normal conditions will be quite sufficient to obviate abuses. A nation which allows itself to be plunged into a war has already lost command of its destinies, and must not be expected to maintain the control over its currency. *Inter arma silent leges* : only a quack scheme can pretend to be proof against the shocks of war-fever—which my scheme does not pretend.

Now for the self-liquidation of the bills that are discounted. The only point that has to be settled is to know what could induce the drawer of a bill not to liquidate it against its maturity, or how he could possibly help liquidating it. He pays the official rate of discount, which cannot be much lower, or much higher, than the current rate of interest. He is likely to obtain more favourable terms if he finances his business with long-period loans. Such loans cannot be secured in a period of rising prices, but when the market is moderately steady, and so long as discounts are readily obtainable, the necessity to draw bills is much limited. As a matter of fact the draft is an instrument which becomes almost useless under a stable currency. My scheme provides that paper money shall only be issued against bills, so that the Bank must always have among its assets as many bills as will amount to the total quantity of notes issued. If the course of events should prove that the business world finds it more convenient to dispense with discounts and pay cash, a way would be found to convert the credits into long-term loans, after which drafts would only be had recourse to in junctures of a momentary shortage of cash, when they would be necessary to prevent a depression.

"In one of its exuberant moods the commercial community could demand the creation of fresh currency as fast as it liked, continually forcing up prices . . ." opines Mr. Withers. But what could create such a mood ? Nothing except the certainty that prices are going to rise. Under the traditional gold standard, the object of the worship of the critical, this certainty was produced as often as new gold-fields were discovered, and from it was born the desire to anticipate profits by getting out as much cash as might be procured. Hence the scramble for discounts. No stretch of my imagination can make me discover an occurrence likely to bring about the same tendency when gold has ceased to act as a seducer, save war and an orgy of destruction—and I refuse to consider means for preserving a stable standard in a time of war and madness.

Like all the critics of his type, Mr. Withers showers encomiums on the Bank of England for managing the currency. He does not seem aware of the fact that this managing always went in the direction which the currency was forced into by the stimulus of extraneous influences. Vigorously as he denounces war-time inflation, which he charges exclusively to the ignorance and cowardice of governments, he has not one word of censure upon the inflation again and again encouraged by the attitude of the Bank of England. These are the closing statements of his chapter on the *Pre-War Monetary System* :

Balances at the Bank of England could be, and were, expanded at moments of exceptional demand to any extent that the Bank of England thought prudent. When demands arose for expansion of credit, and the ordinary banks found that their proportion of cash to liabilities was becoming too low, all that they had to do was to call in part of their loans, which they habitually advanced to the London bill-brokers as their second line of reserve. The bill-brokers, in order to repay these loans, went to the Bank of England to borrow, and it lent them money with which they paid off their loans to the other banks. So the balances at the Bank of England of the other banks were increased and the cash basis of our credit system expanded with extraordinary ease by a few entries in the Bank of England's books. This easy elasticity, which made our monetary system an almost perfect machine when worked by bankers for financing our production and trade, lent itself only too well to abuse by politicians, when a great war called for sacrifice, but the politicians found it simpler to invite us to make money out of our country's need.

Within the twenty years preceding the war the English currency, in terms of commodities, was depreciated by some 30 per cent., thanks to the managing of the Bank and the belief in the soundness of the gold standard theory. (From 1896 to 1914 the purchasing power of the income of Consols was reduced from 139 to 100. See Keynes, *Tract*, p. 15.) 30 per cent.—and by Mr. Keynes's figures I ought to say 40 : is that the extent that "the Bank of England thought prudent" ? What did the Bank derive the substance of its thinking from ; what were its criteria ? So long as gold flowed into its reserves, an average inflation of 2 per cent. a year made no difference and no impression. If more gold, if an indefinite amount of gold had flowed in, that "indefinite expansion" which we have heard him deprecating in such moving tones, would undoubtedly have been approved of by severe Mr. Withers. If controlling and managing is the pride and the excellence of the Bank of England, why does it depend on "demands arising for expansion" ? Why does it not create demand when there is too little or none of it, and why does it not bridle demand when it becomes rampant ?

Our quotation is particularly significant by what it evades. Very beautifully it sets forth the service rendered to expanding business "for financing production and trade" ; but very artfully it drops the curtain over the scene which would exhibit production strangled, manufacturers and merchants made bankrupt, workers turned out of their employment by the million, when, taking the hint from some foreign (or native) mandarins—pigtailed Chinamen or top-hatted Wall Street gentlemen—the Bank of England set out to " manage " the depression which was bound to succeed the boom which it had managed. I am sure the Governor and managers of the Bank of England have always been everything that prudence and integrity could demand ; but I refuse to believe that a better system could not be contrived : a true gold standard, not subject to the prudent judgment of a small body of individuals, but as automatic as any natural organism, which knows how to keep its temperature and pulse constant.

I here insert a paragraph to demonstrate the main fallacy of the scheme of Mr. Lowenfeld in his book *Money in Fetters* (London, 1924). As a preliminary it may be stated that the plan is borrowed, along with most of its arguments, from the writings of the Hamburg banker Friedrich Bendixen, which the interested reader will find very adequately summarized and interpreted in Karl Elster : *Die Seele des Geldes*. It is rather unfortunate that Mr. Lowenfeld has not confessed his indebtedness, of which he need not have been ashamed, because it certainly is something of a distinction to have understood the merits of Bendixen's work. Bendixen contends that the German Reichsbank, for the last twenty-five years before the war, conducted the creation of money on the principle of the self-liquidating bill of exchange, as advocated by Sir Edward Holden, and as described by Mr. Lowenfeld in these terms (p. 88) :

The following should be the exclusive rule for putting currency into circulation and for withdrawing it : (1) The Currency Issuing Bank shall, without restrictions as to amounts, discount at market rates with legal tender money all trade bills offered to it for discount, provided that these bills are guaranteed by a bank of standing which certifies them to result directly out of a goods purchase destined for re-sale. (2) The issuing bank shall at all times re-sell bills in its possession at slightly above cost. No legal tender money shall be put into circulation or withdrawn from it in any other way.

Mr. Lowenfeld endorses the plea for the stabilization of the purchasing power of money: indeed, stabilization is the only aim and object of his effort. He specifically supports his scheme by saying that it was practised in many countries. ("Currency has in the past been put into circulation in this way in many countries, and this well-understood practice would be continued.") However, as he does not imply that it brought stability in its wake, we are led to inquire after the new means added to the insufficient old ones in Mr. Lowenfeld's plan. He would eliminate the use of gold. But that does not solve the difficulty. Bendixen shows—though not very convincingly—that the Reichsbank did not issue notes to suit its gold reserve, but rather bought (or sold) gold to suit its note issue, maintaining that all this dealing in gold had no influence on the movement of prices. All the more necessary it becomes, then, to meet this doubt: the main drift of the movement of the German currency under the scheme was one of a fairly steady rise of prices; if the gold cover clause, which is supposed to inhibit inflationary tendencies, does not suffice to produce results, what is to produce them when this clause is removed? I admit that the reliance on gold was a lure and a snare, and did sometimes lead to an over-issue of currency; but I maintain that the removal of the gold clause would be followed by worse blunders, unless some more powerful barrier is erected in its stead. Mr. Lowenfeld has not considered the action of gold at all, and does not inquire into the cause of the inflation that did take place under the gold standard regulations. Neither does he suggest any change of policy, except in so far as he advocates more caution in accepting bills presented for discount:

At present the issuing bank passes currency into circulation in a similar way, but discounts credit instruments issued by the Government and by others, without paying any attention to the purpose for which currency so issued is used; moreover, bills are sold only when it suits the bank's convenience and the rate of discount charged is fixed by a bank committee. Under the new conditions *it must be charged at the ruling market rate.*

"The ruling market rate": it is here that I and Mr. Lowenfeld must part company. The market rate of interest goes up when and because there is a heavy demand for money. Every merchant knows that this state of affairs will result in a rise of prices, so that every merchant who has his wits about him will strain his means to the utmost in order to buy at once. The consequence is that prices do leap up, and the sums for which bills of exchange are drawn become inflated. Hence the provision that money shall only be issued against bills constitutes no safeguard against inflation, *because inflation begins before the cash-money is created.* Here we have the vice of Bendixen's "classical money" conception and of Mr. Lowenfeld's scheme. It is ignorance of the nature and ways of interest that has misled the devisers of these theories. The problem of interest has got to be solved before a true monetary reform can be effected.[1]

[1] Mr. Lowenfeld wants his issuing bank, which he conceives as a "Government office," to counteract tendencies making for a "deliberate rigging of prices," by which he means "pushing prices out of their natural development." It needs no special acumen to realize that the forces tending to achieve this must be at work incessantly. By adopting the market rate as its rate of discount, the issuing bank would merely give in to the whims of the Stock Exchange and the machinations of the speculators. In order to counteract these tendencies, surely the only way—apart from sheer inaction, which is my method—would be to move the rate in the direction opposed to the trend of the market rate. Does not "counteract" signify acting in the opposite direction? How can you counteract by acting in concert with the very thing to be counteracted?

5. The Currency Standard and State Loans

In discussing the means, as proposed by the reformers, to issue and withdraw cash in a managed currency system, I insisted that these means are not practicable, because they clash with the forces governing the situations in which they would have to be applied. The three main methods recommended are :

(1) Discount policy : the discount to follow, or if possible to outstrip, the movement of the price level. By my interpretation the rate of discount would have to be moved in the inverse direction : it must run counter to the price movement.

(2) State loan policy : State bonds to be sold on a rising market, to be purchased on a falling market. My solution would invert the procedure.

(3) Fiscal policy : taxes to be raised as a check to inflation, and to be reduced to counteract deflation. By my reading of the case the reverse would be the right course.

Before I proceed I must reiterate my conviction that all this managing is wrong. In stating that one way is right while the other is wrong, I simply mean to say that the former would be the right way if managing were necessary.

It is not easy to prove by mere logical reasoning that an argument is correct. I am going to furnish experimental proofs in support of my theory. Indeed, the world, for these ten years, has been one vast laboratory for the testing of monetary theories by experiment.

Mr. Withers has supplied me with an account of the English experiment, which I here reproduce (p. 139) :

> For this and other reasons it seemed that the volume of outstanding Treasury Bills should somehow be reduced, and with this intention a funding loan was offered in June 1919. All the organization and energy which had been so successfully developed during the war for encouraging subscribers to war loans to lend as much as they possibly could and as much more as they could borrow from

their bankers was put into commission to secure the success of the funding loan, but this success was conspicuous by its absence. Already the country was decidedly anxious about the scale of Government expenditure, and investors had become imbued with the belief that the only way to make the Government economize was to refuse to provide it with money. They were decidedly sceptical as to whether the sums which they lent would be used for the alleged purpose of reducing the floating debt, and openly expressed the opinion that the Government was almost certain to expend it upon other objects.

This passage is equally significant by the fact which it states—and by its interpretation of the fact. The fact, to be perfectly clear about it, is this : there was an excess of money issued. Over a thousand million of treasury bills were outstanding. These bills were considered as a part of the floating debt. (In this conception of paper money as being a debt of the issuer lies the root of the fallacy. This point, however, cannot be gone into here.) The purpose of the loan was to convert this floating debt into a funded debt, i.e. to wheedle the public out of their treasury bills (bearing no interest : and I wonder how many of the holders of these bills considered themselves as the creditors of the sums printed thereon !) by offering them in exchange interest bearing bonds. It was hoped that by this means it would be possible to reduce the volume of money in circulation, and so to arrest the rise of prices. The attempt failed. (By the way : if treasury bills were a part of the floating debt, what sense would there be in converting this debt into another State debt ? Those treasury bills were not issued as a loan ; they were an indirect tax on the owners of money and money claims, and what the State obtains through taxation, whether direct or indirect, constitutional or unconstitutional, cannot be considered as its debt.)

The interpretation of the fact given by our expert is at once oddly wrong and curiously right. It credits the public with a degree of wisdom and foresight which no public is capable of, least of all a public whose one ideal still was hanging the Kaiser. The opening paragraph of the chapter from which our quotation is taken accounts much more correctly for the attitude of the public towards the loan. Let me quote a few lines (p. 119) :

> (The Cunliffe Report was not put into practice as soon as the war was ended.) This could not have been expected, even if war's strained excitement had not been followed by an outburst of peace hysteria that was perhaps still more virulent. If we had all been sane and sober as judges, it would have taken some time to adjust our financial system to the circumstances involved by the arrangement of industry on a peace basis. As it was, there was an amazing outburst of political and economic optimism, followed by violent reaction . . . (p. 120). As everybody knows, the war was followed by a spell of delirious trade activity, accompanied by a violent rise in prices, high wages, large profits, and rampant extravagance on the part of both the Government and of the individuals composing the nation.

That was in 1919, at the time when the loan was floated. Now consider the situation created by this very sagacious move. The public, all delirious, saw prices soaring. There was no knowing to what lengths the process would be carried. But everybody had learned to understand that when prices rise money investments are depreciated. What is 5 per cent. interest per annum if depreciation takes 20 or 50 per cent. out of the substance of the principal ? Only perfervid patriotism, or out-and-out idiocy, or superhuman foresight could induce a man to take up the loan ; but the true and downright business man would scorn the notion, and feel insulted if asked to subscribe. As it happened the people who did go in for the article made a very handsome profit : within two years after the issue of the loan the bonds had gained in value and their real value, as measured in commodities, was nearly doubled. The people of foresight had their reward, and for once innocent and unsuspecting patriotism paid. It paid at the expense of the "patria," though, and the lesson from the British funding loan ought to be a warning to those who would have the State interfere in matters of currency. If such a loan does succeed it means that the State has to bear the whole burden of the success and pay enormous profits to the subscribers. The bonds are sold for depreciated money with the express purpose of making money appreciate again. At first sight it would seem as if the seller must gain in this bargain, for does not he get the appreciating money? Yes, indeed, only he may not keep such money, he may not buy things with it, he must destroy it. How else could the loan diminish the volume of outstanding paper money? When the money is destroyed the bonds straight begin to appreciate, and the consequence is that the debtor has to pay interest in much more substantial money, and finally to redeem dear what he sold cheap. Fortunately loans offered in these circumstances cannot succeed, because the trend of the economic forces is opposed to them.

In what respect is Mr. Withers's interpretation right? It is right in so far as it proceeds from the assumption that the money raised for the State by a loan cannot really be withdrawn from circulation, but will be spent, circulated, again by the State. The reformers' schemes could produce the desired effect only to the extent that the money collected through the loan or through taxation is not spent again, but destroyed, by the Treasury. Gesell is the only one, I believe, to say explicitly that it would have to be so dealt with. I do not count the English Professor Cannan among the reformers; but he is quoted by Mr. Withers to this effect (p. 153):

> When the scales at last fall from the eyes of the people of Europe, groaning under the rise of prices, they will no longer cry to their Government, "Hang the profiteers," but "Burn your paper money and go on burning it till it will buy as much gold as it used to do."

It is quite impossible that such a course should ever be followed, whether it be recommended by a radical revolutionary or a rabid reactionary. A sane and knowing person cannot conceive the thing. How is the money for the bonfire to be collected? Suppose it were done by taxation: the State squeezes it out of its taxpayers. The money thereupon is burned, while the huge debt remains: the taxpayers, shorn of their money, have to be squeezed twice as hard, beyond the bearing point, to produce the means for paying the interest on the National Debt. Suppose it is done by a loan, the yield of the loan being forthwith destroyed. In the proportion that money is destroyed money enhances its value, and the State, by acting up to the principle of our case, would have to contract a new debt, big or small, which would be sure to grow in weight as time went on. This talk of burning does not bear a moment's consideration. Contrariwise, if the Government employs the money secured by the loan in the normal manner, it will compete with private enterprise as a buyer in the market, and far from contributing to arrest the push of prices upward, it would be an agent of inflation. And thus every logical and practical reason goes to show that the aim proclaimed cannot be reached by the method proposed. In theory the case may be made to look plausible enough; but the theory is a fallacy, and therefore practically useless.

The British attempt to put it into practice is not the only one of its kind. Jean Labadié, in his book *Si j'étais ministre des finances*, has this remark on the subject of French loans:

> Inflation is a simple matter; deflation is less so. You know the loan permits only theoretically to send paper money to the pulping machine. For two years (written in 1922) we have been raising loans for this purpose, loudly proclaimed, but we have not succeeded in destroying a single one of the treasury bills in circulation.[1]

[1] On p. 169 I have quoted, in a footnote, a passage from a circular of the American Bankers' Association issued with the purpose of

It must not be imagined that to invert the process would work the cure; when ice has failed as a remedy it is still far from sure that heat will succeed. In a previous chapter I have hinted that the purchase of bonds by the Treasury, rather than the sale, might contribute to check inflation—although I did not neglect to repeat that managing was bad in any case. For argument's sake let us try to realize what effect and what success the redemption of National Debt would have had in the England of 1919.

There were plenty of State bonds outstanding, and they were to be had cheap, extra cheap. Whereas the offer of newly created bonds by the Treasury must have further reduced their value, the purchase of bonds by the Treasury could not have failed to enhance their value. Now, an advance in the price of securities at a fixed rate of interest is only possible, whatever the circumstances, at the expense of the dividend-bearing investments. 1919 was the year in which shares were booming in an unprecedented manner. Well, the interference by the Treasury as a purchaser of bonds would have been the way to break the boom of shares. Capitalists, instead of trying to sell their depreciating bonds in order to bid for industrial stocks, would have stuck to their appreciating bonds with the result that demand

ousting silver from the currency. I here repeat this sentence out of it:

"This requires the authorization of $500,000,000 to $1,000,000,000 of new bonds as a basis of circulation."

How is this to be interpreted? The aim was to drive silver money and silver certificates out of the circulation. In order to achieve this end it was necessary to create a willingness on the part of the public to exchange their silver pieces and certificates against bonds. Now bonds would not be bought unless they were made to appear more advantageous than other investments. This condition was fulfilled if the price level was broken and made to sink; the way to bring the desired end about was to restrict credit. Before bonds can be placed and cash withdrawn from circulation through their sale, credit must be strangled in some way or other. As a rule, however, it does not suit anybody to make the attempt when it ought to be made to check inflation: in the first stages of the boom credit is too vigorous to be assailable, and so the issue of State bonds to forestall a boom is a remedy of the same nature as the proverbial "belling of the cat."

for shares was weakened. And that would have sufficed to end inflation.

Now for the Treasury's ways and means to buy bonds. There were two courses that might be attempted : economy and the printing-press. Economy on current expenditure : no purchases of any kind, no new State enterprises, no subsidies for housing, farming, etc., drastic reduction of the departmental staffs, of the army and navy—in a word, the historic Geddes axe at the right time (besides stopping of the mouth of that very ignorant man who was going to search the Germans' pockets). Think of the effect which such methods would have produced. Instead of competing with the makers of price—private enterprise, that is—the State would have withdrawn the ground from under the feet of the profiteers. Demand for labour, for goods, for real capital, and real estate would have slackened, and so everything that was aimed at, and missed, by the loan would have been achieved by the contrary procedure.

However, I grant that this course was not practicable for psychological reasons. Another course was the printing of more treasury bills. If treasury bills were considered as a National Debt, what harm could there be in redeeming one kind of debt (interest paying) by another kind (paying no interest) ? The quantity theory of money of course here raises clamorously the objection that the issue of more paper money would create more inflation. But it is here that the elusive nature of this theory in its crude form will be manifested. The volume of money outstanding is a minor factor ; much more decisive is the velocity of circulation. If we can succeed in reducing this factor, no increase of the volume of money need force up prices. As a matter of fact an addition to the supply of money ought, logically, to weaken the demand for money, and cause its circulation to slow down. Money in this respect must follow the same laws as other commodities. When market goods are offered more abundantly demand takes up an attitude of lofty reserve; their turnover becomes less brisk. And there is no exception to this rule. How is it, then, that an increase of the quantity of money has always been associated with a speeding-up of its pace? It is because a mere addition to the quantity of money does not really increase, but rather diminishes, the actual substance of money. A swelling of the mass of money comes in consequence of increasing speed. Mere mass is only a function of energy, and as energy gathers force, force is subtracted from the mass.

If it were possible to increase the volume of money while diminishing its energy we should be taught by events that such increase does not raise prices. Such increase does occur whenever the energy of the circulation decreases, which happens when the boom is arrested and depression sets in. Then it is that idle money will collect in many places, a great abundance. The mass is swelled by the waning of its energy. But mere mass is unable to produce any movement and any work; when the mass grows it is as though it were absorbing energy into it: energy becomes latent.

The point then to be aimed at would be so to manage the issue of the new notes as to weaken the economic energies at the same time. Energies depend for the display of their powers on direction, a current, a drift, and they are routed when their aim is removed. Inflation is produced by energies operating towards the expansion of certain values; to arrest inflation, energies must be turned off into another direction, i.e. towards the expansion of different values. The values favoured by inflation are the prices of commodities and of the means of production, and inflation must come to an end when other values begin to attract the favour of people. To create demand for a different set of values would be the policy: cause the interest of the speculating public to swerve away from industrial values towards State securities.

We have come round to the point where we can understand the effect produced by the Treasury purchasing bonds of the National Debt with money printed for the purpose. This action creates a new demand for securities, which begin to attract the interest of speculators. Those who but yesterday were endeavouring to sell bonds and acquire shares to-day will be looking out for bonds and ready to offer shares for sale. At once the tension of the energies will be relieved, the wind is taken out of the sails of the boom, which means that inflation is ended. The money issued by the Treasury in buying bonds will not go to the purchase of shares, or of real capital, or of goods, because all these articles have ceased to attract, and, this being so, prices will not rise.

The situation so arising would be a rather disconcerting one. How are the people to employ the money realized by the sale of their bonds ? Whereas the first sellers were happy to invest it in shares and real property, later sellers will shrink from such a venture. Where, then, are they going to place their money ? Money, to be sure, promises to appreciate, and so to treasurize it may seem a good investment. However, bonds are an even better one : they appreciate in a double ratio, in price and in purchasing power, and so they become so much sought after that they cease to be offered for sale. Here the Treasury's action would have to come to a natural close.

Enter now the after-effects. A very considerable quantity of money has been newly issued through the action : where has it gone to ? We need not inquire for particulars ; we know that it must be there, in the hands of speculators. (Anyone who does not spend or invest the money he receives is a speculator.) They will not keep it indefinitely, and, since they shrink from buying, they will try to dispose of it by lending it or depositing it at the savings bank. This move marks the outbreak of a crisis : the super-abundance of cash becomes manifest, and what can be the consequence but depreciation of money ? All of a sudden money will rush by the savings banks, and out of the deposits, to buy, buy, buy : we are back in the current of inflation once more. Thus the action of the Treasury is revealed as a huge fraud : manufacturing money out of nothing, and robbing the people. If I have said that mass is a function of energy, I must now add that energy is also a function of mass : the increase of the quantity of money is bound to translate itself into energy again, which will set the mass into motion and disperse it. The quantity theory of money reasserts itself and will have its revenge. In the long run mass will tell. Special circumstances may allow money to accumulate without affecting prices—for a certain length of time. A classical instance of this was furnished in Germany on the occasion of the French invasion of the Ruhr district, when for some months prices and the rate of exchange were kept fairly constant in spite of huge new issues of paper money by the Government. The end of the adventure ought still to be remembered, and the final effect of our adventure in debt redemption could not be very much different : after a spell of indecision and a show of triumph, the flood would break in and submerge the land in torrents of ravenous money.

Perhaps the point may best be illustrated by the following argument. State loans are issued at a fixed rate of interest. Now it is obvious that these bonds appreciate when money appreciates, but in a double ratio ; and vice versa. In order to regulate the flow of money it will not do to set money off against values which are affected in the same direction as bank-notes. To preserve the balance bank-notes must be exchanged for values which are on the opposite end of the see-saw, rising in price when money sinks. Industrial shares would do beautifully, if they were not subject to special, individual, influences. Bills of exchange are suitable for the purpose, and so is gold. Shares, bills, and gold are the antagonists of money : they represent goods. But securities at a fixed rate of interest are on the same side as money. When money depreciates everybody would be glad to exchange it for things which appreciate; but nobody will care to part with it for an article which depreciates even more heavily than itself. This, however, is the case with bonds: they not only lose in price, so that they represent a smaller sum of money, but the price is paid in depreciated money. It is natural, therefore, that it is impossible to redeem bank-notes at par, through the supply of bonds. And again, when money appreciates many would be willing enough to buy bank-notes, but not if asked to surrender values which appreciate even more promisingly, which is the case with bonds.

The reformers who propose this impossible method are naturally forced to call in the help of special managers and of the State. A natural method can dispense with these aids. Let the Bank of Issue sell freely at par, against its bank-notes, those of its assets which are demanded by the public when money depreciates: gold, bills of exchange; let it buy freely, and at par, against bank-notes those articles which are offered by the public when money appreciates: gold and bills of exchange (bills represent goods, and in buying bills the bank really buys goods which are thereby saved from depreciation). There is no managing required, no spoon-feeding of the market, and no purgatives. Indeed, this managing, whether old style as recommended by Mr. Withers, or new style as advocated by Mr. Keynes, is a euphemism for physicking: feeling the pulse, taking the temperature, examining the tongue and excretions of a patient in a poor way, and requiring to be maintained by doses apportioned by an apothecary's art. A currency that has to be managed is a sick currency. Better not know that it is diseased than all this administering to its sickness.

No kind of interference by the State can right a dislocated currency. The State is powerless to remedy an evil which it has caused. Therefore let us keep the State out of the monetary system. Money is an instrument of business. The State's office is to rule not to transact business and meddle with our business machinery. It is almost pathetic to find the reformist writers on matters of currency unanimous in appealing to the State to do the right thing. Anti-socialists agree with socialists in entrusting to the government the setting up and the control of the money machine. I suppose this is a manifestation of the trend of thinking to which we have been brought by evolutionist conceptions. And quite certain I am that it is an expression of their distrust in their own schemes : only the helpless and the weak appeal to the State.

Mr. Withers is a representative of the opposing tendency, at least in so far as the State is concerned. Rather than to the State he would have the money machine entrusted to the "uncontrolled management" by a private institution. In another respect, however, he does not differ materially from the reformers : he, too, wants a managed currency, although his pretensions are rather more modest, seeing that he wants the currency managed in the direction in which it would naturally tend to go, so that by his theory the managers would make inflation when everything else makes for inflation, and deflation when everything makes for deflation. That is the acme of wisdom, and a method which cannot fail. Indeed, it is the method that has been in vogue, and to which the world has owed most of the fluctuation which has taken place. Without this management by bankers, which only cumulated and aggravated natural tendencies, there would have been vastly more stability. Our investigations into the successes achieved by the wisdom of bankers allowed to manage and control credit has not, then, led to very satisfactory or encouraging conclusions, so that we are forced to demand that the bankers shall be kept out of the currency system as much as the State. We cannot trust to their "thinking" this or that "prudent" or not prudent.[1] We want all the thinking necessary to produce good results done in the devising and constructing of the system, which must be a perfectly automatic one. In presenting the idea of the interest standard of currency I do not claim to have carried the necessary thinking to its perfection. But neither do I ask for believers in my scheme. This book is a challenger. I ask for critics, and shall hail as brother seekers after light those who will take up my subject and carry it forward, nearer and nearer to the point of perfection at which it can and will be put into practice.

[1] Besides, they are busy enough minding their own private interests; it is criminal levity to rely on their also minding the public welfare.

EPILOGUE

TIRED reader, you have passed through the ordeal. Many a time, as I laboured over and dragged you through the intricacies of the theories of the currency reformers, was I reminded of Kipling's ingenious "Just-So Story" of the making of Armadillos. You remember how Stickly-Prickly Hedgehog and Slow-and-Solid Tortoise twisted about and muddled up the rules which Mother Jaguar had given to her inexperienced youngster. Not that I would insinuate that you are the innocent from the banks of the turbid Amazon River. My meaning is that what, in the course of the last years, has been written and said on our subject is of a nature to puzzle the inquirer in much the same way as Painted Jaguar was puzzled. Those schemes are contradictory among themselves and in themselves, which makes it impossible to understand them. In order to clear the ground for my solution I was forced to expose the errors of my predecessors; for to be sure about what can be done you have to know what cannot. But I am somewhat afraid that you may feel like Jaguar when he laments: "And now you come and tell me something I *can* understand, and it makes me more mixy than before." The story, you remember, goes on to show how threatened Hedgehog and precarious Tortoise are saved by combining into a well-adapted and unassailable armadillo, all in the teeth of Mother Jaguar's insistence that "a Hedgehog is a Hedgehog, and can't be anything but a Hedgehog; and a Tortoise is a Tortoise, and can never be anything else." There are people who, in the strength of "old experience," swear that the gold standard can never be anything else but what it has always been known to be. Yet things do change, do they not? The trick of Stickly-Prickly and Slow-and-Solid consisted in the former learning to swim and the latter to curl up. It is much the same with my scheme. It combines two creatures of old standing, whose original advantages have outgrown their time and turned into drawbacks, into one new and living thing, of which it may be said that "it isn't a Hedgehog, and it isn't a Tortoise. It's a little of both, and I don't know its proper name." I have not found the proper name for my new animal, that is my chief trouble. The inflexible historic gold standard and the all-too-changeable rate of interest between them combine and make the adaptable yet rigorous interest standard. The name, I suppose, does not matter particularly; but, indeed, I could have wished for a more comprehensive one, to express the very important fact that there is a combination of forces. For a standard to be the thing it must be a value with regard to another quantity. Gold standard by itself was not enough; interest standard by itself is not enough. But I must leave it at that for the present.

Now what, after all, does my creation amount to? Is the innovation proposed worth the effort of this treatise and your prolonged study? I have asked myself the question more than once when I was impressed with the feeling that I was elaborating a string of truisms. "Keep the rate of discount stable after having established it in its right place, and leave everything else to find its own place in the scale of values:" that is the whole of my scheme. Considering the actual state of affairs in my own country of Switzerland one might be tempted to say: what more do you wish for? In an official survey of the Swiss National Bank's record for 1924 occurs this statement: "The rate of discount did not undergo any alteration in the course of the year. The directors of the Bank are particularly pleased that a rise of the discount rate could be dispensed with, as it has *always been the aim of the National Bank to maintain as stable a rate as might be possible*." (Look at the record of 1914, quoted above, p. 140.) Regarding gold, the policy that has been applied fulfils almost completely the provisions of my scheme. "The National Bank has renounced its former practice of drawing to its reserves the gold coins received at the public counters. In spite of this, or may be rather owing to this, an increasing influx of gold coins has been manifesting itself. Gold coins which have lain in hoards are being returned to the Bank." (What grist to my mill!) "It would seem that in proportion as the note issue is reducing itself it will become expedient also to reduce, in a cautious manner, the gold reserve."

The last utterance proves, as do many others which I need not quote, that the directors of the Bank have not fully understood the case. They want to manage, and imagine they can manage, gold. At a time when the metal presses back to the reserve, they propose to sell gold out of the reserve: cheapen the gold which is already too cheap. This is what managing a currency results in. And they talk of the happy possibility of re-establishing the full gold standard law with a fixed price for gold and convertibility of the notes. Obviously they fail to realize how much more desirable the present, unlawful, state of affairs—where enforced experiment has landed them—is than anything ever known of before. Indeed, the Swiss currency has never yet been more stable than during the years 1923 and 1924. Then why not leave good alone, as men who know the right thing when they see it would do? The purpose of my book is to try and understand what is. It promises no great step forward nor move upward—taking for granted that the more stable a currency is, the better is it. It just points to the successful, though perhaps unconsciously made, experiment and says : you have found the way for yourselves ; therefore persist in it.[1]

Many will deride my effort ; a string of truisms. But would you prefer a string of brilliant errors such as you get in the books of the reformers ? True progress moves by almost imperceptible degrees. What is, is not wrong, but right. How else could it have come into existence ? A true innovation can never do more, I believe, than establish the rights of *something that has brought itself about*, behind men's backs, so to speak. There can be nothing spectacular about it. But on the other hand it does not necessitate any awkward speculations concerning the "necessary period of transition." For my scheme, no nationalization of banking and of wholesale trading in staple commodities, as by the scheme of Mr. Lloyd, which English socialists favour ; none of Silvio Gesell's revolutionary attempts ; none of Professor Fisher's subtle dealings in gold. Extreme moderation is the keynote of my scheme. It was, however, not consciously planned to be moderate. On the contrary, I started rather ambitiously ; but as I advanced and saw through the mist of misconception enveloping the problem, I threw overboard piece after piece those imposing requisites of statistical computation, of managing, of nationalizing, of organization, and international agreements, which make up the panoply of the reformers. And I learned to respect, to honour, to accept the two great factors which have always been the body and soul of monetary systems : the precious metals and interest. What most convinces me of the soundness of my solution is the fact that it has turned out so moderate, so conciliating, so insignificant, and so well tuned to the mood of the times.

[1] Nothing is more true than the opening sentence of John Stuart Mill's *Principles of Political Economy* : " In every department of human affairs, Practice long precedes Science."